HEART OF THE DRAGON

THE ORACLE

THE SAGA OF SHANGALA
BOOK ONE

peter man

booklife prize 2018

critic's report

"The novel's literary style is creative, informative, and at times offbeat, with instances of breaking the fourth wall .... The breadth of knowledge and content Man integrates into the narrative (art history, world geography, mythology, and literature) is ambitious and often captivating. This genre-spanning novel is rich, profound, and--for some readers--potentially overwhelming .... The author's care in crafting the plot and character dynamics--and the juxtaposing of the ancient and the futuristic--are apparent throughout the work .... Shifts and transitions between time periods can be blurred, but the meaningful connections Man draws between the various characters' histories and global events are both enjoyable and educational."

(Booklife is a division of Publishers Weekly)

Heart of the Dragon

The Oracle

Canadian Edition: Sep 2018
First Edition Published 6 July 2018

Publisher email: stoneman.1eye@gmail.com

Book webpage: https://heart-of-dragon.weebly.com/

Social Media:

https://www.quora.com/profile/Peter-Man-18

https://www.facebook.com/stoneman.1eye/

https://twitter.com/PeterMa69468653

Printed by IngramSpark

ISBN-978-1-9994019-0-0 (sc)
ISBN-978-1-9994019-1-7 (e)

Contents

Author's Note

For readers just seeking diversion, I suggest that one should just skip this and plunge right in. The story however has many layers and threads. For those who want to discover hidden treasures, solve mysteries, and explore new ideas, the following may help them gain a basic understanding of some of the language issues.

There are basically two ways Chinese names are transliterated—the old way and the new way. Chinese names before 1982 were generally transliterated into English based on how the names were pronounced in various Chinese dialects. The problem with that system is, not only are there many regional dialects which themselves are not standardized, the English transliterations are sometimes inadequate as well as inaccurate.

Rather than going into great details about how English speakers may get confused, and therefore completely confuse the reader in the process, let us just say that certain names used in the story, for example Chiang Kai-shek, are based on old transliterations because of long usage. The modern spelling would be Jiang Jieshi, which is how a Mandarin speaker pronounces the name, but the old name has been around for a long time, so we'll stick with that.

After 1982, the Chinese government standardized the transliteration of Chinese into the Latin alphabet. It is called Romanization. The result looks like English but it is not English. For example, we used to say that the last Chinese dynasty was the Ching dynasty, but with the modern system, it is now the Qing dynasty, and English speakers have a hard time pronouncing the word Qing. To help the English speaker, we can mentally substitute the Latin *Q* and *C* with *Ts*. Hence, Qing becomes Tsing, and Cao becomes Tsao. *Ch* in Latin is somewhat similar to *Ch* in English.

The pronunciation of Latin vowels is different from English as well. The Latin *e* is approximately uh in English. The word for river in Chinese is *he*, which is approximately huh. The Chinese word for dragon is *long*, which is approximately loong. The word for charcoal is *tan*, which is approximately taan. *Zhou* sounds like joe, *you* sounds like yo, *shi* sounds like shih, *su* sounds like sue, *xu* sounds like she, *xue* sounds like shueh, *xi* sounds like si, and *xie* sounds like shieh. One does not need to be an expert in Latin alphabets to enjoy the story, but it helps to know that there is a difference.

The reader should not however construe any of this as official or complete. What we have here are very superficial and simple examples of how one can use the more familiar English pronunciations to mimic the Latin. Those who are interested can obviously find many sources where they can learn proper Mandarin pronunciations. The Confucian Institute is one of them.

While everyone knows that Mandarin is the official spoken language of China, it is an old term that means 'court language.' The modern term is now Putonghua (普通话; pu-tong-hua), which translates to 'generally accepted language' or simply common language. The pronunciation based on Latin spelling is called pinyin (拼音; pin-yin), which means spelled-phonetic. Pinyin is the standardized pronunciation for Putonghua. It is also a method for computer input of Chinese characters using an English keyboard. In the story, we always use Pinyin to indicate the pronunciation for Chinese words unless otherwise indicated.

With a few specific exceptions, most pronunciation of Chinese words has no blending. For example, Shanghai is pronounced shang-hai, not shang-ghai. Some English and European language speakers make this mistake.

At a few places in the story, we will indicate certain words to be pronounced in dialect or old dialect. The reason is that, foreign tribes have ruled northern China and in particular Beijing for a large part of Chinese history. Beijing dialect, which evolved into Mandarin or court language because of Beijing's position of being China's capital for the past six hundred years, is closer to the northeastern dialects, while the dialects of the rest of the country have much older roots. For example the word 'ten-thousand' (万) in Putonghua is *wan* (sounds like waan), but in Cantonese is *man* (sounds like maan). The word very likely was pronounced *man* in ancient times. Similarly, the barbarian tribe known nowadays as Xiongnu (匈奴) was very likely pronounced *hun-nu* in ancient dialect. That is why scholars have been trying to connect the Xiongnu with the Huns.

The Chinese writing is an integral part of the story and explanations are embedded in the plot. In order to illustrate how modern Chinese characters have evolved from the ancient pictograms known as Oracle Bone script, we have included some of these words in the body of the story.

There are in fact two sets of modern Chinese characters known as Traditional script and Simplified script. The official written language in China is the Simplified script. Taiwan, Hong Kong, and Macau still use the Traditional script. Some Simplified characters have shed pictorial elements and therefore do not have pictorial relationships with the original pictograms. For example, the word for Han (as in ethnic Han) in Simplified form is (汉), but in Traditional form is (漢). The Traditional form of Han is a direct derivative of the original Oracle Bone pictogram. The explanation is inside the body of the story.

Whenever we include Chinese characters in the body of the text, they will always be in Simplified form, unless in some cases where the Traditional form is needed to illustrate the pictorial connections. We will indicate the words with the following format: Han (Simp: 汉; Trad: 漢). There is in fact a large set of characters which are identical for both Traditional and Simplified forms. For Oracle Bone glyphs, we will use the abbreviation OB, and we will use BR to indicate that the glyph is a Bronze script, which existed during the late Shang and early Zhou period.

Another issue is the naming convention. Without getting into historical naming conventions, which is quite complicated, most modern Chinese family names are represented by a single Chinese character, with only a few family names that are double characters (for example, Sima). The given names are usually one or two characters.

Each Chinese character is monosyllabic and transliterated into a monosyllabic Latin word. For example, Dr. Sun Yat-sen is old English convention. The first word *Sun* would be the family name; the given names *Yat-sen* are transliterated into two English monosyllabic words sometimes hyphenated, with the last word not capitalized. In modern Chinese convention, the name would become Dr. Sun Yixian, combining the two words of the given name. This may create problems demonstrated below but we will adhere to the modern convention.

Chinese place names usually combine all the characters in Latinized form which could be confusing. For example, Henan should be He'nan, river-south, not Hen'an; Yanan should be Yan'an, extended-peace, not Ya'nan. Xian should be Xi'an, west-peace, not Xian (pronounced as a single word). If you see a person's name written as Chen Xian, you will not know if his given name is Xi-an, two words, or Xian, one word. Despite the potential confusion, we will not deviate from the existing convention, but we will add a word separator when it is helpful to do so.

As luck will have it, two adjacent Chinese provinces have exactly the same Romanized form. If you are going to Xi'an to see the Terra Cotta soldiers, you will be going to Shaanxi (陕西; pass-west) province, which is right next to Shanxi (山西; mountain-west) province. Shaanxi is the old transliteration used in this case to distinguish it from Shanxi. Strictly following Romanization rules would make both provinces Shanxi.

Most Chinese names and place names have meanings. Romanization means keeping the pronunciation but losing the meaning. Our story tries to give meaning to names when appropriate. For example, rather than using the Romanized name Bailangshan, we will keep the name White Wolf Mountain because the plot ties this location to a character by the name of White Wolf.

The historical Shang queen Fu Hao becomes Queen Zia in our story. This transformation is not haphazard and the explanation is inside the body of the story. We have likewise changed King Wu Ding (Martial IV) into King Wotan, and the barbarians Gui-Fang (Land of Spirits) into the Gotts. English readers may find it easier to relate to these Westernized names. In cases where we must keep the traditional name, we will provide a short explanation.

Throughout the story, we will use BCE (Before Common Era) to replace the older term of BC (Before Christ), and CE (Common Era) to replace AD (Anno Domini). For dates without indication of CE or BCE, we can assume that it is CE.

For those readers who are interested in Chinese history, Chinese languages, and Chinese culture, this story provides a teaser for the real thing. The intention is to pique the interest of the reader. The story covers many fascinating facts and events besides what comes out of China. Prepare to be amazed!

CHAPTER 1

Everything is a Lie

"Everything is a lie ..." – Santos Didymas, Sage of Shangala

Our tale begins not with an opening crawl of bright words on dark background followed by epic orchestral fanfare. Neither is there an introduction by a beautiful enchantress with a melodious voice over fairy tale music accompaniment. We have no wizened wizard's gravelly voice to set the tone for a most unusual adventure. And we refuse to use that morose monotone to recite another moronic monologue describing a post-apocalyptic world. We're left with either a poor emulation of Homer's invocation of the Muses, or a shameless plagiarism of Dickensian rhetoric, "It was the best of times, it was the worst of times," neither very palatable to the modern reader. We therefore decide to launch directly into the crux and climax of the unlikely, unfortunate, and unfathomable events that befell upon one teenage girl by the name of Victoria Solana.

The saga began or rather came to an end with the tinkle, tankle, and the tintinnabulation of clanking steel. It woke Victoria from her dark dreams. She was greeted by a headache of biblical proportions. She felt as if all her bones had become disjointed. A scorching fire raged underneath her skin. The sweet smell of burnt flesh filled the air. During that momentary state of confusion when one was jolted out of deep sleep, she noticed that something had gone terribly wrong. The first sentence that came through her clenched teeth was, "Where is my unicorn?"

Victoria Solana was the most normal and ordinary girl you could ever know. She had lived in a typical suburban neighbourhood in Toronto all her life. As an only child, she was doted upon by her wonderful but undistinguished parents. Shortly before this tale began, Victoria had just completed a blissfully happy and perfectly uneventful childhood. It was her thirteenth birthday. Her only adventures in her non-adventurous life up until now had been her recent visit to the Bethune House Memorial at Gravenhurst, Muskoka, and the exhilarating expectation of getting braces for her pearly white teeth, the perceived imperfections of which, by any and all standards of observation, could only be described as imperceptible. But she must have braces because her BFFs were having them.

Victoria had three faint freckle-like moles on her face. One was high on her left cheek where beauty moles were usually located. Another one was near the left corner of her lip, which was supposed to signify she would be a connoisseur of fine cuisine. The last one was right at the centre of her chin below her lips, which, by the mystical science of molesophy, meant that she was a loving and considerate person.

Victoria shared many interests with her mommy, such as arts and craft, cooking, singing, and meticulously grooming herself. She had a close relationship with her daddy as well and shared some of his interests in science, classical music, board games, and spectator sports. While her daddy loved to watch soccer, Victoria was much more of a baseball fan.

Victoria was an avid fan of the *Amazing Randi* and *Penn & Teller*. She watched most of their shows—those rated PG-13—and read many of their books. She taught herself how to do some of their tricks and learned the lesson never to believe in magic.

Naturally, Victoria became adept in prestidigitation and misdirection, and she would always have a few tricks up her sleeve. She also had an inexplicable fascination in stargazing. For that, her daddy bought her a telescope and spent many happy evenings with her peering at the constellations in the sky.

Although Victoria's parents encouraged her to read the holy books of various creeds to seek her own faith, they generally expressed an informed non-dogmatic mixture of atheist and agnostic views. They called themselves God-free—they were free of the gods, but they were not godless. Hence, Victoria followed her parents on the so-called non-intrusive humanist path simply described as 'Live and Let Live.'

For reasons unknown to Victoria and it was never explained to her satisfaction, her parents did not allow her to partake in any sports, physical games, or competitions. It was as if her parents were afraid that she would draw too much attention. They simply hated to be in the limelight.

Despite living a sedentary lifestyle all her life, Victoria was amazingly healthy. She could not remember ever having visited a doctor or a dentist. In fact, she didn't even remember the last time she took cough medicine or used a Band-Aid. Victoria had been extremely lucky. Life had been all too harmlessly and graciously Canadian for Victoria, and Victoria was as Canadian a Canadian could ever be, ranking way up there with the beaver, Canada goose, maple syrup, ice hockey, and saying sorry when someone steps on your toe.

But without warning and in the blink of an eye, Victoria's life was turned upside down and inside out. She lost everyone in her family and everything she held dear. She could no longer be sure who she was. Her past life became a puff of smoke, suspended precariously in the air, to be dispersed into nothingness with a wave of the hand. By and large, she was like an ant on a Möbius strip, at a loss even as to which dimension she belonged to. Victoria was lucky to still have her two best friends, but she would soon be separated from them by a continent and an ocean.

Never having travelled beyond Muskoka, about an hour and a half's drive from home, Victoria ended up half way across the world in a strange land, trying to survive among people who spoke a foreign language. Meanwhile, she had to evade sinister assassins who were stalking her every step of the way, cheated death in a fiery car accident on the highway, and stared down the barrel of a pistol aimed point blank at her face. She jumped off a plane mid-flight without a parachute, was in a high-speed car chase on one of the world's longest sea bridges, and was trapped inside the car when it plunged into the sea. She dangled from the window at the top of the tallest building in China, wrestled with a hungry tiger, and killed three grown men using only one arm. All this happened to Victoria Solana because, without being aware of it herself, she held one of the greatest secrets of the universe.

"Stop exaggerating," groaned Victoria as she struggled to get up, "just tell the story the way it happened, please."

Dear Lord! This has never happened in the history of storytelling. The heroine of this tale has just made a direct request for how her story should be told. Given that she is the protagonist, she does have the prerogative. For those readers who may find the pompous and hifalutin lexicon of this literary fabrication discombobulating, it means Victoria Solana is the boss, and we do what the boss says.

So let us begin again, with less poetic license and with minimal embellishments, to describe what happened to Victoria Solana and why she found herself rolling on the ground looking for her unicorn. As we all know, the unicorn is a mythical beast; it has no place in a factual story.

CHAPTER 2

The Accident

It was a dark and slippery night. There are many such nights in Canada during the winter, especially on the 22nd day of December, the day with the least amount of daylight in the year. Victoria and her parents, Michael and Angela, were on their way home after having enjoyed a scrumptious and sumptuous birthday dinner at their favourite Chinese restaurant, the *Dragon's Lair*. Victoria loved Chinese food. She learned all about Chinese cuisine by fanatically following the long-running cooking shows of the indefatigable Yan Can Cook. One might think that perhaps it was because Victoria was ethnic Chinese, but it was not so. Michael and Angela Solana adopted Victoria when she was about one year old. She had no memory at all about China, her country of birth.

Michael and Angela were both born in the Ukraine. They had known each other since they were kids. They got married soon after they left school, and enjoyed a loving marriage ever since. When they had the chance, they decided to immigrate to Canada and leave their secret past behind. They even changed their names after settling in their new home in Toronto. Their life in Canada was effortlessly perfect except for one imperfection. They wanted to have children but Angela was barren and Michael was sterile.

In a casual conversation with a Chinese art-and-antique dealer by the name of David Huang, Michael discovered the possibility of adopting from China. That option had become available to

Canadians since China implemented the so-called *One-Child Policy.*

What prompted China to make such a policy? The story began way back in 1979. China was at a watershed moment in its history. The U.S. and China normalized diplomatic relations on the first day of the year, more than six years after Nixon went to China with an olive branch. The *Gang of Four*, the perpetrators of the power struggle behind the decade long Chinese Cultural Revolution, was securely behind bars. The new leader of China, Deng Xiaoping (邓小平), one of the few remaining first generation top tier leaders of the Chinese Communist Party (CCP), visited the U.S. in the same year. The American public with the help of the enthusiastic mainstream media embraced this diminutive giant just as they adored the giant panda bears.

In order to clear the obstacles for the modernization reforms, Deng Xiaoping said he didn't care whether the cat was black or white as long as it caught mice. In a few words, he gave an invaluable advice to the Chinese people to be pragmatic rather than dogmatic. Who was it that said "Dogma is a bitch"?

For the benefits of anyone who might want to understand the secret of China's success, Deng Xiaoping openly shared his formula in a famous one-liner: "China would cross the river by feeling for the stepping-stones."

In modern language, we may describe Deng's method as the *3XI.* The first 'X' is eXperience. Always apply knowledge and experience to develop new economic ideas. The second 'X' is eXperiment. Always start small-scale experiments to test the new ideas. The third 'X' is eXpansion. Duplicate and expand once the experiments provide positive results, all the while monitoring closely, getting frequent feedbacks, and making appropriate adjustments. The 'I' is Innovation. Any successful development will always enjoy a high rate of growth in the beginning, slowing down and going flat over time, eventually sliding downwards into decrepitude. In order to rejuvenate the economy, the market needs innovation. New dynamic economies must replace old moribund ones, and new paradigms must adapt to the constant changes in market conditions.

With ideological baggage out of the way, Deng Xiaoping opened the doors wide for the great Land of the Orient to do business with America and the rest of the world. It was however not without risk. The West led by the U.S. wrote the modern rules of the game—rules such as Bretton Woods, Plaza Accord, Washington Consensus, WTO, Petrodollar, World Bank, and IMF, etc. They were cold and calculated capitalist rules. When the West wanted something to be rule-based, they meant based on their rules.

In addition, the West as the rule-maker could change the rules any time, ignore the rules, or unilaterally abrogate the rules if it was expedient to do so. The West had all the game pieces while China had none. Those pieces were not free. China must pay to play. Furthermore, the West could create game pieces out of thin air whenever they wanted. The West also had a special rule known as 'All Options Are on the Table,' which would usually cause weaker players to concede right away, considering the alternative would be at least a black eye, a bloody nose, and several dislodged teeth accompanied by a loss of face. For the impertinent and recalcitrant, it could be much worse. It was a game the West did not play to lose.

Deng figured he didn't have many choices. China had very little resources except a billion penurious people, most of them half-literate peasants. China had to play the West's game if it wanted to unleash the productive and creative capabilities of its populace, so that the country and its people could realize its full potential. As a big country, China could endure an occasional insult, even absorb a punch now and then, but it could also learn to play the game and over the long haul improve the lives of its people. That is the ultimate goal of the Chinese government. It does not matter what colour cats others may think they are keeping. As the quantum physicists well know, Shrödinger's cats are both black and white at the same time.

Deng sat down at the table, started the clock, and the game was afoot. China was prepared to open up its heart and mind to accept Western myths, Western mores, and Western methods, but always with a twist called Chinese characteristics. By adding a little english on that Chinese spin and cultivating that certain inexplicable je ne sais quoi, they have been causing sinophiles to fawn and sinophobes to foam. Anyways, tourists and investors started pouring into China. The future for China began to look like a glass of champagne, bubbling with hope and excitement.

What many people didn't realize was that Canada had already normalized relations with China since 1970, when the iconic Liberal Prime Minister Pierre Trudeau was leading the country. As an example of Canada's humanitarian approach, in 1980, the country offered without publicity to treat Deng Xiaoping's paraplegic son Deng Pufang (邓朴方), a victim of the Red Guards during the Cultural Revolution. Canada, as the close ally of its southern neighbour, most certainly helped in the final rapprochement between China and the U.S.

The result of Deng's gambit was, by 2017, China had 700 million people lifted out of poverty, with the expectation to eliminate extreme poverty by 2020. Meanwhile, the national GDP of China rose from USD 190 Billion dollars in 1980 to USD 13.5 Trillion dollars in 2017, elevating China to the second largest economy in the world behind the U.S. These are indisputable achievements. The good cat has been catching mice.

As the gates of the Central Kingdom—a more accurate translation of China than Middle Kingdom—opened in 1979,

China was experiencing rapid growth of her burgeoning population of one billion, which had doubled in only one generation of Communist rule. Chinese culture always favoured large families. Better healthcare and nutrition helped more babies survive childhood to become reproducing adults. Such unbridled growth in population could lead to only one result—human disaster.

Heeding the advice of several studies from the West, including the very popular report *The Limits to Growth*, Chinese officials began in 1979 to implement the so-called *One-Child Policy*. It was a more humane way than allowing merciless nature to reduce population by disease, starvation, and war. There were exceptions for people living in the countryside and for ethnic minorities, and of course, for those who knew how to skirt the rules. The Chinese government would revisit and reconsider the policy after one generation pending results.

The plan worked well enough and the population growth slowed down. While this inevitably created some new problems, the alternative might have been much worse. One of such problems became a boon for Michael and Angela.

At the time when the *One-Child Policy* came into effect, many Chinese families still held the traditional opinion that they must have a male offspring to perpetuate the family name. Some baby girls were not registered or simply hidden until they became adults; others were abandoned. A few fortunate ones found their way to the orphanages and some eventually ended up in North America with their adoptive families.

As for Michael and Angela, they received the kind assistance of the art-and-antique dealer David Huang, and came home after a short trip to China with a beautiful baby girl in their arms. That Chinese baby would grow up to become the attractive teenage girl at the back of the car this winter evening, busily texting her friends from her smartphone.

Victoria had long silky black hair, almond-shaped eyes, and the egg-shaped face of the classic beauty of northern China, who would be usually described as active, strong-willed, and at times even martial, like the famous Mulan depicted by Disney. At age thirteen, Victoria was just such a northern rosebud about to burst into bloom.

"Honey, give your eyes a rest," said Angela. "You've been staring at the phone all evening. Father, please talk to your daughter."

"I'm sending food pictures to the *Weird Sisters*," said Victoria, ignoring her mom's request. "The *Lion's Head* cooked in chicken soup and the *Chairman Mao Braised Pork* were so awesome. I want to make the girls drool."

Although Victoria knew that she was originally from China, Chinese food would be the only connection she had with her country of birth. She knew very little about China other than what she could learn from the odd Hollywood movie and the mainstream news. As far as Victoria was concerned, she was one hundred percent Canadian.

Victoria had been best friends with her neighbours and classmates Jackie and Bella since they were babies. The three girls formed a secret society, which they named the *Weird Sisters*. Angela and Michael did not socialize at all. The *Weird Sisters* was the only social life Victoria had, and she treasured it. Victoria thought of her parents as fiercely private eccentrics and didn't think much of it. One didn't get to choose parents.

It was a dark and slippery night. Although traffic was light, the freezing rain made driving on the highway most treacherous. Michael had a dark and troubling thought in his mind. He would rather not get involved in any arguments between the two persons he loved most in the world.

"I have to concentrate on the road," said Michael, driving cautiously at a speed barely below the limit. "We'll discuss this when we get home."

Michael decided that delay was the best policy.

Without warning, the shadow of a leviathan lunged from behind and swallowed whole the Solana family. It was a forty-foot—nay—a fifty-foot truck, zipping by at what seemed like a hundred klicks per second. All it needed was the slightest brush and the dastardly deed was done. Michael could do nothing but watched in horror as his car spun towards the concrete median barrier, while the culprit vanished into the distance like the fleeting shadow of a ghost.

It would take less than one second for the car to careen into the concrete barrier, but it felt like an eternity as the world spun in front of Victoria's eyes. No one had time to react. Victoria wanted to scream but could emit only silence. All she could hear was the dreamy music of *Bach's Air on the G String* playing on the car's sound system. When the initial shock passed, Victoria felt strangely calm and peaceful. It was as if she was inside an out-of-control space capsule, spinning its way to infinity. Having no means to change her inevitable fate, the only thing she could do was to accept her impending doom.

Michael and Angela were not surprised that their time had come. They were almost expecting it. Their only regret was not having the time to tell Victoria that they loved her, and to tell Victoria the truth about her background.

CHAPTER 3

The Unbreakable China Doll

The mighty eagle spread her wings and soared high in the azure sky. She flew over the snowy tops of lofty mountains, glided across the golden dunes of a sprawling desert, and circled above verdant hills and sleepy vales. She could see below her a winding river, churning and gurgling muddy waters as it raced into the ocean to the east. She turned towards the sun that was just rising from the horizon. The majestic bird of prey imagined that she was a dragon as she cast a sinewy shadow on the idyllic villages that dotted the meadows and the grassy knolls below. As she crawled across the languid sky, a harpoon-arrow whizzed by. Then another just missed her by a hair … and another.

The magnificent eagle beat her mighty wings to stir up a turbulent storm. She flew into a column of rain cloud to evade the missiles. It was so dark she could not see where she was heading. Her body was soaked. Water flooded her eyes and blurred her vision. Blue and green lightning flashed around her followed by colourful explosions like New Year's Eve fireworks. She did not know what hit her but it felt as if 1.21 Gigawatts of electricity had singed every neuron in her body. Victoria Solana regained consciousness with the freezing rain beating on her face. For as long as she could remember, Victoria had this recurring dream in which she was a big black eagle.

The car had slammed into the concrete blocks on the passenger side. It was impossible to tell if Angela had been thrown out or buried in the twisted steel. The violent deformation of the chassis had sheared off the rear window. The freezing rain was pouring through this opening. Victoria felt the aftershock of the violent impact from the top of her head to the tip of her toes. It was a miracle that she didn't suffer any cuts or fractures. Victoria was extremely lucky.

Michael was alive but trapped in his seat, barely saved by the airbag.

Victoria wept in shock when she couldn't find her mother.

"Calm down … listen," Michael struggled to speak but it sounded like a whisper. "Leave now … go home … look for birthmark … in my journal."

"But I can't leave you here," sobbed Victoria, "I'm not leaving."

"Great danger … not accident … avoid police … I'll be fine … go now, run!"

After the collision, lighting for this stretch of the highway went dark. A black SUV had stopped by the shoulder several hundred metres (or yards) away. Its emergency lights were flashing.

Victoria squeezed through the opening and climbed over the concrete block. There was no opposing traffic. She turned to look at her father who mouthed, "I love you. Go." She wiped her tears and stole off into the darkness.

When Victoria heard the siren of a police car arriving, she turned to check on the rescue of her father. The freezing rain had stopped. Eerie green lightning was flashing in the sky while deep thunder rolled in the distance. Then horrors of horrors—a streak of lightning snaked down from the sky and hit a bullseye on the wreckage. The pile of twisted steel immediately burst into flames followed by a huge explosion. The lightning and the fireball lit up Victoria's face in quick succession. In the blink of an eye, she had become an orphan all over again. Although in shock, Victoria remembered what her father told her. She turned around and ran, and she ran for dear life.

Victoria didn't know how far she ran or for how long. She had no memory of how she ran into a half deserted plaza, how she found a parked taxi with the driver inside taking a coffee break, how her debris covered face almost startled the driver out of his skin, and how he drove her home after she adamantly refused to be taken to a hospital.

Victoria took a hot shower to wash away the shivers and the horrors of the evening. She tried to work out whom she could turn to for help. All her life, she had gotten used to the fact that she had no immediate relatives—not a single cousin, not an aunt or uncle, not a grandparent. It had always been that way. For Victoria, it was normal. Apparently, her parents became orphans after a disaster they did not want to talk about. Victoria had questions but she did not delve into her parents' past. All parents for their own reasons keep secrets from their children. Victoria knew that her parents had suffered trauma. She was prepared to let them tell her when they were ready. Now her questions would never be answered. For the moment, Victoria's biggest questions were who was trying to kill her and why?

Chapter 4

Riddle me This

Although still suffering from the trauma of the evening, Victoria remembered her father's instruction to look for her birthmark in his journal. When Victoria was very young, she learned from her mom that she had a birthmark on her back. Skin blemishes bothered Victoria, but she felt better when her mom told her that birthmarks usually disappeared over time. In fact, she felt very much better when she forgot about it altogether. Out of sight, out of mind.

Victoria got in front of the mirror and twisted around to see if she could find any unsightly spots on her back. To her surprise, she noticed a tiny red heart-shaped mark on her lower back.

This thing looked more like a tattoo than a birthmark. But why would someone give her a tattoo when she was a baby? Maybe her biological parents used that to identify their newborn daughter. Why then did they abandon her if they were afraid to lose her? Victoria had questions but no answers.

Down in the basement study where Michael loved to read in quiet solitude, Victoria searched for her daddy's journal. The room was packed wall-to-wall with books, maps, and notes, some of which were strewn on the carpet here, there, and everywhere.

Fortunately, Victoria knew where her father had hidden his precious notebook. He had kept it in a safe place inside the gap between the top of the bookcase and the low ceiling. It's a place Victoria couldn't reach when she was small and mischievous. She stepped on a stool and shone a flashlight into the gap, and voila! She found it.

The notebook which Michael called his journal was an ancient parchment manuscript. It was supposed to be full of arcane and esoteric puzzles. The front cover had a line drawing of the belly of a tortoise shell. Right at the centre-line of the shell were three strange drawings aligned vertically.

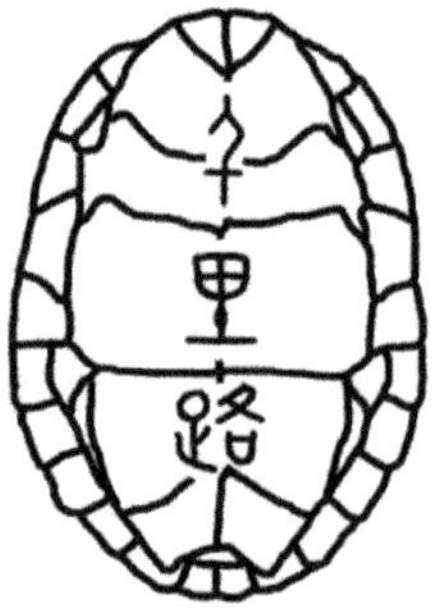

For clarity, the three glyphs are displayed horizontally here from left to right: 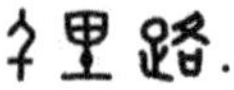. On the inside page of the cover were these three symbols also written vertically: 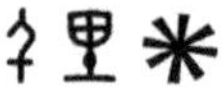.

Fantastic sketches and puzzling ancient glyphs filled the journal's pages. It reminded Victoria of the enigmatic and undecipherable Voynich Manuscript she had read about in the news. Victoria riffled through the booklet to look for her birthmark. It did not take long at all.

Strange looking glyphs filled the page on the left. One of the glyphs was Victoria's heart tattoo. The page on the right had a star chart showing stars and constellations that definitely looked alien. At the centre of the star chart was the drawing of a flower. Victoria was quite certain she had seen the flower before, but she couldn't quite remember where. In the margins were illustrations of otherworldly flora and fauna, including the face of an animal with a single horn. A small arrow pointed from the 'heart' glyph to a line of scribbled comment: "To learn secret, decode secret."

Victoria was confounded. The puzzle did not say what secret to decode. How could anyone decipher a secret with no clues? It's impossible! Victoria gave way to frustration and rummaged wildly through the mountain of books and piles of files. It was hopeless.

After regaining calm, Victoria thought of a lesson she learned from her father—when looking for a hidden object, it is important to use the brain rather than the eye or by brute force, and one should never depend on luck or hope without thinking and taking action.

Michael had introduced Victoria to the game of crossword. She was in fact rather good at solving all kinds of puzzles. "How would daddy leave a clue for me? Would it have something to do with my tattoo?" Victoria thought. "No, anyone can tell that it's a heart. It's too obvious."

Victoria read the clue again.

"To learn secret, decode secret."

But what secret could it be? Then it dawned on Victoria that the secret could be the word 'secret' itself. To learn the secret, decode the word 'secret.'

Victoria knew that the answer for the crossword clue 'secret' was *sub rosa*, the Latin expression for 'under the rose.' This must be it. The secret of her biological parents was hidden under a rose.

She paced around the room carefully scanning the books. The first section on the bookcase was religion, the next was archaeology, and the third was art and music, followed by science and technology. On the shelf below were the Greek and Latin classics, next were *Gibbon's Decline and Fall of the Roman*

Empire, Joseph Needham's *Science and Civilization in China*, and history books on Alexander the Great, Hannibal, Julius Caesar, the Duke of Marlborough, Frederick the Great, and Napoleon. The next shelf displayed all of Shakespeare's plays and sonnets, and the complete collections of Sherlock Holmes and Agatha Christie. The shelf below was devoted to books and journals written by I. F. Stone, Upton Sinclair, and Gore Vidal. Then Victoria hit the mother lode.

Right by the side of the bookcase was a stack of old books, DVDs, Blu-rays, and CDs. The stack started from the top with a Unitel-Classica High-Definition video production of the Richard Strauss opera *Der Rosenkavalier* on Blu-ray. Next was a Franz Schubert lieder compilation CD. Underneath that was a small booklet. It was a poem by Goethe titled *Heidenröslein*. Another booklet underneath was titled *The Rosicrucian Manifesto*.

Next in the stack were six DVDs. The first one was the Disney classic *Sleeping Beauty*. The second DVD was *Titanic* by James Cameron. The third was the biopic *La Môme* about the French singer Édith Piaf. The fourth movie was the all-time classic *Citizen Kane* by Orson Wells. The fifth was a PBS documentary titled *The Roosevelts*, and the last was a Canadian documentary titled *The Children of Soong Ching Ling*.

All the DVDs sat on top of the bittersweet fable written by Antoine de Saint-Exupéry, *The Little Prince*. Directly underneath was the Shakespearean play *Romeo and Juliet*. Then it was Gertrude Stein's book *Geography and Plays*, followed by a hardcover book titled *The Diet of Worms*. The next tome was a history book titled *The House of Plantagenet*, and finally at the bottom of the heap—the giveaway clue—Umberto Eco's bestselling suspense novel, *The Name of The Rose*.

To the unsuspecting, this might have been just a random pile of books, movies, and music, but to Victoria, the stack screamed ‘roses’ to her.

Victoria checked carefully through the stack. Still nothing! Victoria had great confidence that her father did not place all these rose-related items in one stack by accident. She must go under the carpet. Victoria found a sheathed Swiss army knife in her father’s mess. She severed the carpet from the floor with the dagger after moving the stack off to one side. True as the night is long, hidden underneath the carpet was a secret compartment. Victoria removed the cover to reveal a small box. It was made of a very light but sturdy material. The box had a symbol on its lid.

CHAPTER 5

Sub Rosa

No doubt, some readers may wonder how the stack of books, movies, and music are all rose-related. The following is a short explanation:

The first rose is *Der Rosenkavalier* or *The Rose Knight*, the most popular opera by Richard Strauss. The opera interestingly features a male role sung by a female singer.

The song *Heidenröslein* is the lead song in the Schubert lieder compilation CD. It is based on Goethe's poem of the same name. *Heidenröslein* means 'little rose on the heath.'

The Rosicrucian Manifesto is a pamphlet describing an ancient secret society known as *The Brotherhood of the Rose-Cross.*

In the Disney classic *Sleeping Beauty*, the fairy godmothers call Princess Aurora by the name of Briar Rose while she is growing up in the forest.

Rose is the name of the heroine in the blockbuster movie *Titanic.*

La Môme tells the life story of the French singer Édith Piaf, whose most famous song is *La Vie en Rose.*

In the classic film *Citizen Kane*, Charles Foster Kane utters with his dying breath, "Rosebud."

The Roosevelts is a documentary about the Roosevelt clan featuring its two Presidents and one First Lady. The name Roosevelt means 'rose field' in Dutch.

The Children of Soong Ching Ling is a Canadian documentary about the wife of Dr. Sun Yat-sen (孙逸仙; sun-yi-xian) and her lifelong work helping orphans in China. Dr. Sun Yat-sen was the pioneering revolutionary who founded the first republican government of China. All Chinese regardless of political persuasion revere him. Soong Ching-ling (宋庆龄) supported the Communist revolution. She is a beloved historical figure in China. Rosamond was Madam Sun's name when she attended Wesleyan College in the U.S. Rosamond means 'pure rose.'

In *The Little Prince*, the rose is the reason the prince leaves his home planet to learn the meaning of life and love.

In *Romeo and Juliet*, Juliet says these immortal words, "What's in a name? That which we call a rose by any other name would smell as sweet."

There is a poem titled *Sacred Emily* in Gertrude Stein's book *Geography and Plays* with the following memorable line: "Rose is a rose is a rose is a rose."

The Diet of Worms is not about modern cuisine or sustainable diet, nor is it a philosophical contemplation on mortality. It is about an imperial assembly (diet) of the Holy Roman Empire at the German city of Worms in 1521 CE, which resulted in the excommunication of Martin Luther. This led to religious, political, and social upheaval of great consequences. The seal of Martin Luther is a white rose with a cross in the centre.

From the House of Plantagenet sprouted the House of Lancaster represented by the red heraldic rose and the House of York represented by the white heraldic rose. The struggle between the two houses for the English throne is known as The War of the Roses.

The last line of *The Name of the Rose* says, "The rose of the past remains only in its name; all we possess are empty names."

By now, it should be painfully obvious that Victoria Solana was in fact an extraordinary teenage girl. Victoria had lived all her life trying to act average because her parents did not want any attention. Now we know why. They were in hiding. Her parents were dead because someone finally tracked them down.

Back in Michael's study, Victoria opened the box that would lead her to the answers. The first thing she saw was another strange symbol on the underside of the lid.

Inside the box was a folded letter from Michael. It was written in his very neat cursive handwriting.

Honey, by the time you read this, your mom and I may already be gone. Don't be sad. Be strong and be resourceful. You are an extraordinary girl. Your mom and I have sworn to protect you until you are reunited with your biological parents. Without you, we would have died long ago. We're blessed with the years we have spent with you. There are powerful people who want to take you from us. This box is prepared in case we have failed to protect you. Inside the box, you will see some cash and a three-piece jade pendant. You had this pendant when we first saw you. Always keep it with you. You must find David Huang, the man who led us to you. He has a company called Solvicta Antiquities. There is a symbol at the back of this note. Use it to identify him. You should keep my journal. It is a rare treasure I got from a blind fortune-teller. You may find it useful someday. Leave the house as soon as possible. Be careful and stay alive. We love you, Dad and Mom.

p.s. I love you! Mom

A teardrop fell on the letter as Victoria finished reading. She turned the letter over to find a symbol on the back:

Victoria folded up the letter and put it in her knapsack. Inside the box were three wads of cash. They were one hundred dollar bills in U.S., Canadian, and Chinese currencies.

The jade pendant was underneath the money. It was made of three jade pieces connected in a tier by a red string and tied in intricate Chinese knots into a necklace. The jade pieces were white nephrite of great age. The top two pieces were in the shape of an arch.

The first piece had a symbol carved on one face. On the reverse was a string of symbols.

The second piece had similarly carvings on both faces.

The last jade piece at the bottom was a ring. Carved on the exterior was a pattern of an animal face with a horn at the centre of its head.

On the back of the ring were two symbols.

Victoria pulled the necklace over her head, and stuffed the pendant inside her shirt. She did not find any answer in the box. She discovered more mysteries instead. She grabbed a few essentials, stuffed them in her knapsack, and walked out of the life she had ever known.

Christmas would be here in two days. Most of the homes in the neighbourhood were festively adorned with Christmas lights and garden ornaments. It was supposed to evoke happiness and excitement, but all Victoria could feel was sadness and emptiness. She was all alone in the world. Where should she go from here?

As Victoria walked past the homes of Jackie and Bella, she felt compelled to bid farewell to her best friends. She couldn't just disappear from their lives. But what should she tell them? Would telling them about what happened endanger them and their families? Jackie and Bella were only a few months older than Victoria. Could they keep Victoria's secrets?

CHAPTER 6

The Weird Sisters

Victoria lived about five minutes' walk away from Jackie and Bella, both of whom came from affluent families living in large homes. Jackie's father was a successful importer of olive oil, feta cheese, and ouzo liqueur from Greece. His family had emigrated from Athens several generations ago. Bella's parents were solidly Canadian of English background with a long history in the country.

Victoria always thought that Bella had cool parents. Michael and Angela worked at average jobs, perhaps deliberately. Bella's father on the other hand was the professor of Cosmology at the University of Waterloo, and her mother was a cryptologist working at a government agency with an acronym. Conversation between Bella's parents often revolved around whimsical topics such as Higgs Boson, Strange Matter, Yang-Mills Theory, Lie Groups, Galois Theory, Twin Prime Conjecture, Riemann Hypothesis, Fermi's Paradox, and other imponderables.

Bella's parents were iconoclasts who did not subscribe to the common wisdom repeated ad nauseam that a single beautiful line of equation would describe the universe in its entirety ad aeternum. Victoria once had the good fortune to listen in on a discussion between Bella's parents on the imperfect human perception of reality.

Nature, according to Bella's father, was like an elephant being described by a number of blind scientists who could only examine the animal with their hands from a single position. While the blind men might guess the beauty of symmetry for the elephant's left and right sides, it would be a thankless task for them to reconcile the ear with the tusk, or to fit all the bewildering elements of the elephant into a single mathematical symmetry.

Does this perception of reality have anything to do with the strange glyph on the inside of the treasure box? The drawing on the underside of the lid clearly showed a hand touching an elephant.

After launching a few well-aimed pebbles at Jackie's bedroom window, Victoria was able to get her out of the house without alerting anyone else. Jackie then summoned Bella to an emergency *Weird Sisters* meeting with a late night phone call.

Victoria and Jackie were huddling in the backyard when Bella opened the door to the back entrance and beckoned them to get in.

"Let us do the meeting outside," whispered Victoria. "I don't want anyone else to know that I've been here."

Bella wrapped herself in a warm overcoat and came out.

"Girls, something terrible happened tonight," said Victoria, holding the hands of her best friends. "I can't say too much except I'll have to leave immediately and I don't know when I'll see you both again."

Victoria's words shocked Bella and Jackie into stunned silence.

"I have to go to China to look for my biological parents," said Victoria. "I'm going to miss you both very much."

"What about your parents here?" Bella asked, "What on earth happened?"

"They're dead, and if I don't leave now, I'll probably be dead too," Victoria sobbed softly. "Our car got hit by a truck on the highway and then it exploded after I got out. Before that, my dad told me that it was no accident. Don't let anyone know you have seen me, not even your parents, and definitely not the police. Be very careful about strangers in the area. I have to go now. I can't leave without saying goodbye."

"We're not going to let you go out there all by yourself," Jackie said. "We're the *Weird Sisters*, Vicky. One for all and all for one, remember? You'll have to let us know how we can help you."

"No, I'm sure once the people who killed my parents find out that I'm still alive, they'll be hunting for me," said Victoria. "I mustn't get you two involved. It's too dangerous."

"I know what to do," said the intrepid Jackie, who was brought up on Greek tales of heroism. "Let us set up new chat accounts using fake names. You'll let us know your whereabouts and what help you'll need from us."

"We can use our secret code words," said Bella.

As with all secret societies, the *Weird Sisters* had developed a set of code words such as WAY meant 'Where Are You,' IAH meant 'I'm At Home,' and YSM meant 'You're So Mean.' They're not hard to decipher but it's more fun to chat in codes.

"Let us decide our new accounts and nicknames now," Jackie said. "My account ID will be Atropos, and my nickname will be Aisa. Search accounts for Atropos and if you see the nickname Aisa, that'll be me. Bella, your account ID will be Lachesis, and your nickname will be Lacey. Vicky, your account ID will be Clotho, and your nickname will be Chloe. Everyone got this?"

Victoria and Bella both nodded. The three girls were familiar with the Greek goddesses of Fate because the strange goddesses corresponded to the three witches known as the *Weird Sisters* in Macbeth, which was where the girls got their inspiration for the name of their secret society. Atropos is the inevitable fate, Lachesis is the disposer of lots, and Clotho is the spinner of the threads of life.

Victoria was always intrigued with sisterly trinities. She knew about the *Three Sisters* of the Orion constellation, also known as *Orion's Belt.* There were the three historical Soong sisters who were the most influential women of China in the 20th century. And she was familiar with the three influential global think tanks, the Bilderberg Group, the Trilateral Commission, and the Council on Foreign Relations, coined by some as the *Three Sisters,* which aspired to create a new world order based on American democratic capitalist principles. It was a mystery to Victoria why the 'three sisters' kept popping up in the story of human civilization. But Victoria understood the simple truth that girls everywhere must stick together.

At that very moment, a black SUV sped through the usually quiet neighbourhood in the direction of Victoria's home. Its headlights went off as it passed Bella's house. Victoria knew that the hunt was afoot.

"I have to go now. I'll let you know my whereabouts when I can," said Victoria. "Don't worry. I'll be fine."

The freezing rain started to fall again. Victoria bid farewell to her best friends with the secret *Weird Sisters* three-way finger-lock, then she turned to walk away. While leaving behind the life she had ever known, Victoria tried to savour the sweet sorrow, and not let the moment be lost in time—like tears in rain.

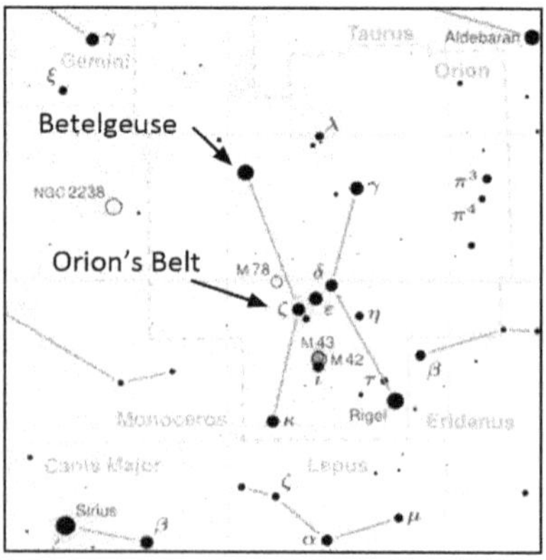

CHAPTER 7

Room 929

After parting with Jackie and Bella, Victoria caught a night bus and went downtown. She figured that she would be less conspicuous in the city and she would be safer among crowds. Victoria stayed up through the night at a 24 hour Tim Hortons, the ubiquitous coffee-and-donut shop in Canada. Other than a couple of homeless people, one of whom had a glass eye that seemed to follow her, Victoria had the place to herself. Using the free Wi-Fi at the coffee shop, Victoria found Solvicta Antiquities' address. The company was nearby.

Christmas was pressing hard upon the city. Festive spirit filled the air. Christmas music and carols were playing everywhere 24/7. The Hudson Bay Department Store on Queen Street as usual wowed pedestrians with its annual Christmas window displays. Yonge Street, purportedly the world's longest street, was festooned with Christmas decorations and fancy lights.

People started appearing on the street while it was still dark. Delivery trucks brought food supplies and newspapers. Soon, well-groomed young men in black suits and fashionably dressed ladies with painted lips rushed into cafés and restaurants for breakfast and coffee. Toronto was coming alive. Victoria polished off her croissant breakfast, crushed the crossword puzzle in the daily newspapers without breaking a sweat, slung the knapsack on her back, and headed out for Solvicta Antiquities.

As Victoria was weaving through the white-collar crowd, she had an eerie feeling that someone was tailing her. She was certain that it was the homeless fellow with the glass eye. That creepy man had stayed in the shadow all night, and when he left the coffee shop while it was still dark, Victoria noticed that he glanced at her with a menacing glare.

Victoria waited for the streetcar, which was what Torontonians called their electric tram, to rumble by, found a gap in the traffic, and shot across the street without looking back. She then raced like the wind until she reached her destination. Victoria had no idea she could do a hundred in ten seconds flat.

Solvicta Antiquities was just a numbered door at the office section of an ancient downtown hotel. There was no company signage. Victoria pressed the doorbell and the door buzzed opened.

The room was cluttered with gaudy Chinese artifacts. Spurious looking certificates covered the walls, authenticating one treasure or another. What seemed out of place were two large Renaissance style oil paintings casually placed against the wall. Both paintings depicted the gruesome scene of two women in the process of cutting off the head of a man with a large sword.

Victoria recognized one of the paintings to be *Judith Beheading Holofernes* by the famous Renaissance master Caravaggio, and the other to be *Judith Slaying Holofernes* by Artemisia Gentileschi, a famous female painter who painted in Caravaggio's style. This extraordinary female artist applied her rare talents to make a powerful statement against men who had abused her and yet enjoyed impunity.

A strategically placed desk commanded the unobstructed view of the room. At the desk was a handsome South Asian man who adamantly denied that the company had anyone by the name of David Huang. Fearing that if she couldn't find the Chinese art dealer, she would have nowhere else to go, Victoria refused to take no for an answer. She was certain that she had come to the right place because she noticed some secret clues hidden in plain sight.

"I'm sorry to bother you," Victoria said, showing the symbol on her father's letter to the man. "There must be a Mr. David Huang here. I have a letter here for Mr. David Huang of Solvicta Antiquities."

"Oh, you mean Mr. Soong," exclaimed the man when he saw the symbol. "But he is in Mumbai and won't be back until the first week of January. Would you like to leave a message?"

Victoria's heart sank. Then she saw something strange. The man slowly put his index finger up to his lips making the hush gesture. Then with the other hand, he lifted a piece of paper with the number 929 on it.

Room 929 was at the end of a long corridor. Whoever rented this room was giving Victoria another secret message that she had found the right person. She was about to press the doorbell when the door opened. Standing in front of her was a middle-aged and bespectacled Chinese man of about the same age as her father, but with salt-and-pepper hair and a greying moustache.

"It's good to see you, Your Highness. I've sent out a search party looking for you," the man said as he bowed respectfully with a dip of his head. "I am David Huang, your humble servant. Sorry about what happened to your parents. Please come inside."

Victoria was puzzled with this 'Your Highness' business.

"It's okay," the man said. "I'm the David Huang you're looking for."

He reassured Victoria by raising his right hand and showing her the gold ring on his index finger. It had the same symbol that was on the back of Michael's letter.

CHAPTER 8

Sign of the Scorpion

A lone eagle floated high in the clear blue sky, enjoying the serene beauty of the world beneath her. With her eagle eyes, she observed a band of nomads emerging from a mountain range that separated the wind swept steppes in the west from the lush green valleys and dark-soiled plains to the east. Their eastward path followed a small river that carried the waters from within the mountain range into the plains below.

Although it was not apparent at this point, the humble muddy stream would soon join with other tributaries to form a major river, the lifeblood of iron tribes that would rewrite the history of China and forever change its character. The roiling waters would then disgorge out of the Dragon's Mouth into the endless ocean hundreds of kilometres to the southeast. The nomads called it the yellow river—not the famous Yellow River, the cradle of the Chinese civilization, but the cradle of peripheral civilizations which would someday become an integral part of China.

The band had no more than three hundred men on horseback with about the same number in women and children. The whole group of nomads looked wild and haggard. Their horses were in a similarly wretched state. A pack of hungry wolfdogs straggled along with them. These canines were more wolves than dogs, and were not too far removed in temperament from their feral cousins. So far, they had faithfully taken part in their human masters' arduous trek over the mountains, but they were at the end of their rope and could break ranks at any time.

Upon reaching the foot of the mountain, the group decided to pitch tent and rest. They were too tired, too hungry, and too wounded to go any further. Almost all the men wore bandages. The few that were not limping or too weak from bleeding had to set up the tents with the women while the children took care of the horses. Some women and men went off to forage. The sun was setting, but this being summer time, they would still enjoy another hour or two of light.

The younger and stronger men quickly set up a large tent made of sheepskin and wolf pelts. Several older women dragged a dying man on a sled into the tent. Three young men promptly erected a large bronze spear at the entrance. This was the symbol of the chief. After all the helpers had left the tent, a woman came out and summoned the three young men to enter.

The chief was lying on a cot. Beside his head was the tattered and bloodied banner of the Tuman (头曼: tou-man), which was what the people called their tribal chief. The Tuman was badly wounded. He was on his last legs.

The three disheartened young men stood by the deathbed. They were the evidence that the tribe had suffered a devastating defeat on the battlefield. The one standing closest to the dying chief had a blood soaked cloth wrapped around his head. His right hand was also heavily bandaged. He might have lost a finger or two, maybe even three. The second young man was limping around with the help of a makeshift crutch. His left thigh and his right arm were also bandaged. The young man standing farthest appeared to be without serious wounds, though his tunic was caked with dried blood. The chief had called his sons to receive final instructions.

The woman turned to leave the tent so the men could talk in private.

"Lupa, wait," said the Tuman. "You are my children's wet nurse and their mother's handmaiden. You have been with the family all this time. Now that your mistress is with the wolf spirits, and I will join her soon, you are the only person my sons will listen to. Please stay and be witness."

Lupa preferred not to be involved in politics, but she couldn't disobey the Tuman. She retired to a dark corner and kept quiet.

"White Wolf, North Wolf, Lone Wolf," the Tuman said, "put your hands in mine."

"Baba," the two wounded sons called their father in unison, but the third whispered, "Tuman, my lord."

"Lone Wolf, call me Baba," ordered the Tuman, "I gave you your life."

Both White Wolf and North wolf turned to glare at Lone Wolf with disdain.

"Yes, Baba," said Lone Wolf.

"I am sorry, Lone Wolf," the Tuman said, "that against your counsel I went to war with that despicable father murderer."

White Wolf and North Wolf turned and spat as they heard the mention of he who slaughtered the Eastern Wolf tribes.

"Now that the Hunnu has destroyed the Eastern Wolf confederation, our own tribe of ten thousand bows and uncountable sheep are no more," said the Tuman. "Curse the Confederate Tangri (Chief) for his greed. Curse him for causing death and destruction to my family and my tribe. I will curse him again when I meet him after death. I have only a short time to live. I can do no more.

"Promise, Lone Wolf," the dying Tuman said, "that you will stand by your brothers and help our tribe return to glory."

"Yes Baba, you will recover and we will survive," said Lone Wolf, keeping a brave front. "We will be strong and numerous as in the days past, our wolf-howl will strike terror into our enemies' hearts, and the Tuman banner will fly in the wind again. I swear to fight side-by-side with my brothers White Wolf and North Wolf as you command."

"Good," said the chief, who then turned to his other sons. "And will you two swear to support Lone Wolf like true brothers?"

"Why of course, Baba," said White Wolf. "Ever since we saved Lone Wolf from the wild animals, we have kept him with us like a part of the family, even giving him our tribe name, he is like a brother, and we'll always treat him like a brother."

"Even if I make him Tuman?"

It was a tradition of the Wolf Tribes for the chief to pass the Tuman banner to the most capable son regardless of age or rank. Still, none of the young men was prepared for this shocker, least of all Lone Wolf.

Three years ago, the Tuman saved Lone Wolf's life during a hunting expedition. He found the young man near death, wrapped in a black shroud and abandoned to die in the wilderness. Lone Wolf was suffering from a strange affliction that caused his skin to turn red hot. The Tuman threw him into an icy stream to quench his fire. He ordered Lupa to nurture the young man back to health, and afterwards kept him as a son, even giving him the name Lone Wolf.

Ever since Lone Wolf recovered, he had been most helpful to the Tuman and his tribe. Although Lone Wolf had advised against joining the Eastern Wolf Confederation's war, when the time came for the slaughter to begin, he fought most valiantly, and he singlehandedly retrieved the wounded Tuman who was surrounded by enemy troops.

"Baba, I am your eldest son. I am more experienced and the people respect me," said White Wolf. "Besides, Lone Wolf is not even of our blood. I don't think it's wise."

"Baba, even if White Wolf can't lead because he lost the use of his sword hand," said North Wolf, "I should be your second choice. I can fight again when I heal."

"Baba, Tuman my lord," whispered Lone Wolf, meekly adding to his adoptive brother's obstreperous objections, "I agree with my brothers. Please reconsider."

Ignoring Lone Wolf, the Tuman fixed his gaze on White Wolf and North Wolf.

"Yes, you will heal to fight again, but it's the fighting that got us here," the old man said. "All of us great warriors did not have the wisdom of one Lone Wolf. When you all sneered at him and called him a coward, he did not get angry, but fought beside us. Everyone saw his bravery in battle. No enemy soldiers could stand their ground before him. But even more important, Lone Wolf prepared our escape while we were planning on how to distribute the spoils. We all owe him our lives.

"As for Lone Wolf's foreign blood, remember that the Wolf tribes are made up of different peoples from faraway lands. This makes us strong. Your mother came from beyond the ocean of sand where the sun sets. Her tribe spoke a musical language and worshipped strange gods in the sky. Your mother had grey blue eyes, like a wolf. Your cousin, the girl Wolf Star, has reddish hair, green eyes, and fair skin. What is her blood?

"I want a wise person to become Tuman after I die, to lead whatever is left of us, to make the clan strong again. Lone Wolf was a gift from the gods."

Turning towards Lone Wolf, the Tuman used what strength he had to impress upon the reluctant young man, "Swear, Lone Wolf, that you will not abandon us. Swear that you will be Tuman after I die. Swear that you will bring glory again to our Wolf tribe."

The three young men saw that it was pointless to argue against the deranged rant of a dying man, so they just kept their heads down and their mouths shut.

"Swear," shouted the Tuman with his last breath.

"I swear," whispered Lone Wolf.

The Tuman smiled a smile of satisfaction as he laid down his head and gave up the ghost.

A few curious children who had been peeking into the tent immediately scattered off to spread the news. Within seconds, a chorus of howling began to rise from the camp, soon joined by the wolfdogs. The beloved Tuman was dead. Long Live the Tuman! The last remnants of the once proud tribe sang their laments and their dirges, while a new struggle began for the succession inside the Tuman's tent.

"Lupa," said White Wolf, relying on the support of his most important ally, "Baba was not in his right mind. Surely you must know that."

Lupa did not speak but walked over to the deathbed and put two pieces of jade on her master's eyes. She then turned to stare sternly at White Wolf and North Wolf. "Your father had a good reason for making Lone Wolf Tuman," said Lupa.

Then turning to Lone Wolf, Lupa raised her arm and pointed at the young man. She demanded with a sharp voice, "Lone Wolf, show us your mark."

Lone Wolf hesitantly turned around and lifted his thick long hair to reveal a red tattoo on the back of his neck. It had hitherto been hidden from all except the Tuman and Lupa. At the same time, Lupa lifted up the chief's banner to display the sign of the Tuman. They were a perfect match.

CHAPTER 9

Dragon Heart

The people promptly buried the great chief without pomp and circumstance, as befitting a defeated leader of a band on the run. His sons strangled his favourite wolfdog White Fang and buried it along with their father as his companion in the afterlife. Lupa convened the tribal meeting after the burial, and promptly proclaimed Lone Wolf Tuman. White Wolf and North Wolf begrudgingly kept their mouths shut. They knew they were no match for Lone Wolf even if they were healthy. Hence, Lone Wolf received the old Tuman's great bronze spear and the Tuman banner without bloodshed.

For this solemn occasion that would determine the future of the tribe, Lone Wolf put on his eagle mantle, which he had made by attaching eagle feathers to his black shroud. The new Tuman looked majestic as he addressed his adopted tribe.

"I am alive today because of the old Tuman; may his spirit protect us always," said Lone Wolf. "I have promised him that I will lead the tribe, but I cannot accept the title of Tuman for now."

White Wolf and North Wolf were surprised at what they heard, and their eyes brightened up with the hope that the Tuman position might still be up for grabs.

"The title Tuman means the head (头: tou) of 'ten-thousand' (万: wan; old dialect: man). The scorpion sign of the Tuman is an ancient symbol passed to us together with the title from more than a thousand years ago. The Han (汉) Empire in the south, following the tradition of the previous Qin (秦), Zhou (周), and Shang (商) dynasties, also has adopted the scorpion to mean 'ten-thousand.' It is therefore a much-revered symbol.

"As such, I will not be Tuman or carry the Tuman banner until our tribe has ten thousand bows again. In the meantime, I shall be Chief Lone Wolf to you all, and I shall lead the tribe with the help of my two brothers, the Right Chief White Wolf and the Left Chief North Wolf."

This prompted a wave of murmuring among the tribe members, but most nodded their heads in approval.

"We have to make some immediate decisions," said the new Chief Lone Wolf. "Modred (冒顿: mao-dun; dialect: mo-du), the murderer of his father, wishes to exterminate us."

Everyone spit at the mention of the cursed name.

"His soldiers are following our tracks through the mountains at this very moment," the young chief continued. "They are only slowed down by the threat of an ambush. I suggest that we should keep moving through the woods rather than to descend into the valleys and the plains. In the mountains, we can hide better, and we can ambush our pursuers if the opportunity presents itself.

"The trees are laden with fruits and nuts. We will not go hungry here. The woods are also teeming with deer, reindeer, hares, pheasants, and wild birds; we can hunt them and trap them for food. There are foxes, minks, and sables; we can trade their furs. Our wolfdogs will be happy here. And if we don't stray too far from the streams, we will eat our fill of spawning sturgeons, northern salmons, burbots, and carps. I will teach you how to catch them with nets.

"There is enough pasture for our horses and our herds. Where the land is flat and suitable, we can grow millet. I have saved up some seeds. Finally, we shall worship the golden *Shanbe*, the heavenly beast with the single horn that has led us through the woods to safety. Let us pray that it will guide us through the difficult times ahead."

Lone Wolf had scouted out the area in anticipation of the retreat. He learned that this was the ancestral lands of the Eastern Wolves, where legend said that the golden unicorn known as the *Shanbe* once roamed. Lone Wolf understood how a bit of faith and hope would be good for the tribe's survival.

"I believe that the Hunnu soldiers will not hound us forever. Modred will soon summon them back for other wars. The Hunnu tribe, once our Wolf brothers, became strong in the seventy black peaks of the *Great Green Mountain* (大青山: da-qing-shan) also known as the *Mountain of the Great Canopy* (阴山: yin-shan; shady mountain). Led by the treacherous father-murderer Modred, they have won a great victory over the Eastern Wolf tribes—the Eastern Wolf tribes are no more.

"Modred has taken an uncountable number of slaves and animals. His tribes are laden with mountains of plunder. The blood of the slain flowed like rivers and the earth turned red. Now that the Hunnu has a taste of gore, gold, and glory, they will never satisfy their bloodlust. Their unquenchable thirst will consume them like a scourge.

"The Hunnu under Modred will soon take their iron hooves to the west and the south. They will enjoy military success. But this will be the beginning of their downfall. This is also the beginning of our rise from the ashes. We must try to find as many survivors and stragglers as possible, and we will join forces. Living in the mountains and the woods means we will not be fighting any territorial wars. The people living in the plains call themselves the Berserkers (貉貊: he-mo). They are fierce fighters who wear the skin of the bruins and charge as if possessed by the animal spirits of old. They will not welcome us.

"In the mountains we will toughen ourselves and our children. We will strengthen our sinews and sharpen our skills. We will stay out of war and bloodshed. We will pay tribute to Modred and make alliance with our dispersed brethren hiding in the mountains, both to the north and to the south. We will use all of our energy to grow, until we are strong and numerous enough to ride out of the mountains without fear. Then I will take up the Tuman's banner and ride with you. All who agree say aye."

The whole tribe punched the air with their fists and shouted their approval as one. A few started the familiar wolf howl, the sound of which might have prompted the people of the southern land to call the northern Wolf tribes the Hu (胡 ; northern barbarian).

"We have had a tiring day, let us rest early tonight," said Chief Lone Wolf. "I'll take the first round of sentry duty. We strike camp on first light to move into high ground. This tribal meeting is adjourned in the name of our old Tuman; may his spirit protect us always. Everyone, have a good rest."

The whole tribe rejoiced in the leadership of their young chief. Everyone went to sleep with empty stomachs but with hearts full of hope. They knew that tomorrow would be another difficult day, and that their survival would depend on wise decisions, hard work, tenacity, suffering, and sacrifice, but they believed that as long as they wholeheartedly followed Chief Lone Wolf's leadership, the tribe would survive and be strong again.

Lone Wolf found a perfect sentry spot on a grassy knoll that overlooked the camp. He had sent a few responsible young men to cover the periphery in other directions. Lone Wolf brought his favourite wolfdog Phantom to keep him company. As soon as he sat down, he heard a rustling noise. Phantom did not make a big fuss because both he and Lone Wolf knew someone had been tailing them.

"Wolf Star, show your face," said Lone Wolf. "You really should be resting rather than training for an ambush."

Wolf Star was kinfolk of the old Tuman's wife. Her tribe had migrated slowly over hundreds of years from the west, across the Celestial Mountain (天山: tian-shan), and beyond the ends of the known world. They joined the Wolf tribes when the chieftain's daughter was married to the old Tuman. Because of the disastrous war with the Hunnu, Wolf Star lost everyone from her old tribe except for Lupa. She was just fifteen years old and all alone in the world. Normally, a girl of her age would be married with children, and indeed, Wolf Star was pledged to someone from her old tribe, but that person lost his life in the fight.

Lone Wolf had always been good friends with Wolf Star. When they were younger, they used to spend evenings looking at the stars, talking about everything and anything their innocent hearts desired.

"I want to congratulate you for becoming our chief," said Wolf Star, as she lay down beside Lone Wolf. "We all trust that you will be a great leader. Chief Eolycus; I like the sound of it."

Wolf Star spoke her old tribe's tongue. She had used it to make a special name for Lone Wolf since they became friends. Her name in the old tongue was Lupus.

"That is a heavy burden I don't want," said Lone Wolf. "But I made a promise to the old Tuman. May he rest in peace."

Lone Wolf was in his late teens or early twenties. He found the aroma and body heat of a beautiful young girl so close to him to be quite unbearable. He took a deep breath and created some space between himself and Wolf Star.

"I missed talking to you," said Wolf Star, as she narrowed the distance between them. "We haven't chatted at all after I was pledged."

"I missed chatting with you too," said Lone Wolf. "And by the way, I'm sorry about your loss."

"Everyone has suffered loss," said Wolf Star, as she sat up and came into physical contact with Lone Wolf. "But while we live, let us live."

The intoxicating fragrance of Wolf Star reminded Lone Wolf of a sad happiness hidden in a memory from faraway and long ago. He asked rather out of the blues, "Do you know why you're called Wolf Star?"

Wolf Star shrugged.

"You're called Wolf Star because you're as beautiful as the Wolf Star in the sky," said Lone Wolf. "It is hiding now beneath the southern horizon, but before the snow falls, it will rise again, and it will be brilliant. You will be blessed with the greatest fortunes on earth; but you must be patient."

"How do you know so much?" asked Wolf Star, who could not take her admiring eyes off Lone Wolf. "Do you have a star?"

"Indeed I do," replied Lone Wolf pointing southwards at the long constellation close to the horizon. "See that string of stars? It is the Dragon, and at the chest of the Dragon is a fiery red star. See it? You can't miss it. It is the *Heart of the Dragon*, and it is my star."

On this night, the flaming red Antares of the constellation Scorpius, first known as the Great Fire and the Shang Star, later as the *Heart of the Dragon* in ancient China, was particularly bright and dominant in the sky.

"Your star has brought you good fortune," said Wolf Star, as she leaned over and put her head on Lone Wolf's shoulder. "I hope it burns bright in the sky to keep you safe always."

"Wolf Star, I'll tell you a secret you must promise to keep," said Lone Wolf. He could not fend off Wolf Star's advances, so he had to tell her the truth. "I already have a wife. We also have a baby daughter before we were separated. The time will come when I'll have to leave the tribe and go look for my family. I hope you understand."

Teardrops like sparkling diamonds fell from Wolf Star's eyes. She said nothing as she held on to Lone Wolf.

Antares of Scorpius
(Heart star of Dragon constellation)

CHAPTER 10

Flight of Fancy

Victoria dreamed that she was a black eagle flying high in the clear lavender sky. The land underneath was shrouded in thick sooty smog. She descended in wide circles until she could see the land through the haze. The air down here was pungent and repugnant. Plumes of fire in colours purple, blue, green, orange, and yellow spewed forth from vents in the ground. The green fields had turned into a land of desolation. Huge concrete blocks and long steel bars were broken haphazardly and strewn willy-nilly all over the land, as if the planet had suffered a devastating disaster.

Victoria was a lover of nature and felt disgusted by this disgraceful sight. What catastrophe had caused the destruction of the planet? How could intelligent people be so criminally negligent? Then she saw something falling from the sky. Victoria could see far with her eagle eyes. It was not a meteor from outer space. Victoria was startled by what she saw—this object had arms and legs. It was a young boy plunging in freefall to certain death. She must save him.

Victoria folded her wings and dove after the boy. It was very dangerous because they were so close to the ground. When she finally grabbed the boy, she struggled to open her wings, but it was too late as she and the boy plummeted inexorably towards the long steel shards that protruded from the ground.

The wakeup call woke Victoria from the nightmare. She was safe and sound in a hotel room. David Huang had arranged a suite in the hotel-office complex where she could rest. After experiencing the trauma of losing her parents and having gone through the whole night without sleep, Victoria was at the brink of emotional and physical collapse. She recounted to David Huang the events of the previous evening, including the discovery of the jade pendant, but she was careful to withhold details about the journal, and said nothing about Jackie and Bella. She locked the door after David left and fell into a deep sleep.

By the time Victoria woke, it was past seven in the evening. She freshened up, ordered sushi for dinner, and called David Huang.

David arrived with a travel bag.

"Prepare for an adventure," David said, "we're flying to China on a private jet. We'll have to find your biological parents. But before we go, we need to make some changes for you."

David took out a pair of scissors and a comb. "You're getting a haircut and a disguise, Your Highness."

"Not a chance," Victoria cried. "And stop calling me Your Highness."

"Yes, Your Highness," said David. "But I'm afraid you'll need to have a different appearance … even a different gender. The bad guys will be coming after us. This is necessary if you hope to survive long enough to ever meet your real parents."

"Who are these bad guys?" asked Victoria. "What have we done to deserve their ill will? We have always minded our own business, and always kind to people."

"It's a long story," said David. "On the other hand we have a long journey ahead of us. It's five hours from the Island Airport to Vancouver International Airport, where we'll fuel up and get your papers done. Then it's twelve hours to Shanghai Pudong

International Airport. We'll have lots of time to fill you in. Now, let us get to work."

Aside from being an antique tradesman, it turned out that David Huang was also an illusionist. When he was done, Victoria saw herself in the mirror and couldn't believe that she actually looked like a handsome young boy. David then coached Victoria to speak with a lower pitch, and taught her to act with boyish mannerisms. This wasn't very hard for the extroverted Victoria, as she was a bit of a tomboy at times. When they were ready to leave the hotel, both were satisfied that David was going on a trip with his handsome young son.

David snapped a couple of photos of Victoria, and then took her in a limousine to the Billy Bishop Island Airport, where a rented Gulf Stream jet was waiting. From this point on and at least for the time being, Victoria became Victor, a Chinese Canadian boy on his first trip to China with his filthy rich Chinese dad.

Using the hotel Wi-Fi, Victoria sent a short message to the *Weird Sisters* before setting out: "Chloe calling, IMOK, OMW2PVG, ⚙ KIT."

Victoria, or Victor, was nervous and excited about flying. Although she always dreamed of being an eagle, she had never actually flown before. This was her first time. She would be visiting the mysterious China for the first time, and she would soon meet her biological parents for the first time.

The handsome South Asian man from David's office greeted Victoria as she entered the Gulf Stream jet.

"Welcome aboard," he said. "Would you like something to drink before we take off?"

David Huang entered after Victoria. He flung a large bag on a seat. "Victor, meet Raja," said David, introducing the two for the first time, "Raja, meet Victor. Are we ready for takeoff?"

"Yes. Flight path was approved," said Raja. "Estimated takeoff time is about half an hour. Would you like a drink before I join the pilot? I'm co-piloting."

"Thanks, Raja," said David. "We'll help ourselves."

"Please allow me, sir," Raja insisted, as he started pouring the chilled Baron de Rothschild Brut into a champagne flute. "We have bubbly cider for Master Victor."

Victoria excitedly watched the ascent from the window. It was awesome. She could see the lights of Toronto recede below her to become a city of stars. Soon the space outside became completely black except for the flashing light at the tip of the wing. Then her familiar constellations came into view.

The first constellation Victoria learned to identify was Orion the giant hunter. Michael taught her about the stars that made up the constellation. The most prominent was the fiery red giant Betelgeuse as the giant hunter's right shoulder; some call it Orion's right hand. Belatrix or the Amazon star, the female warrior, was his left shoulder. The most easily recognizable part of the constellation was the triple stars that formed the *Orion's Belt*, sometimes known as the *Three Sisters*. Finally, there were Rigel and Saiph as Orion's feet.

Victoria reminisced about the times her father told her bedtime stories about Orion, how the giant hunted with Diana—the Greeks called her Artemis, sometimes also known as Luna the moon goddess—and threatened to kill all of earth's animals. That did not go down well with Mother Earth, who sent a scorpion to stop Orion. With one sting, the scorpion dispatched the boastful hunter. Zeus, the head of the Greek pantheon placed Orion together with the scorpion up in the sky as two major constellations. The two gods were however implacable foes. To keep the peace, they would never show up in the night sky at the same time. Orion rises in the winter while Scorpius rises in the summer.

When she came out of her galactic wandering, Victoria realized that David Huang had dozed off in his seat. He must have had a tiring day.

Though it was not possible to forget about all the terrible things that had happened the night before, Victoria convinced herself that this was an auspicious beginning of her adventure. She was after all flying in a private jet. This was the meaning of freedom. She was like that big eagle in her recurring dreams. She could spread her wings and fly anywhere she wanted. Then it struck her that in her nightmare, a boy fell from the sky, and now she knew who that boy was. That boy in the dream was her. She had that eerie feeling that something was not right, and that her dream was a warning. But it was too late. She was 15,000 metres (45,000 feet) in the air.

Victoria wanted to warn David, but he had fallen into a deep sleep. He made no response when Victoria nudged him hard with her elbow. David was out cold.

Victoria realized that someone must have laced the drinks with sedatives. She just drank the cider too. No doubt, she'd be knocked out in a minute. How careless of her. Angela had taught her never to accept food or drinks from strangers. And how careless of David. Why did he trust this Raja guy? Now they're doomed.

CHAPTER 11

Freefall

The determined nudging of David by Victoria brought Raja out of the cockpit. "Master Victor, please let Mr. Soong rest," Raja said. "He was very tired today. He must have taken a sleeping pill."

"I'm sorry, but I don't think so," exclaimed Victoria. "Mr. Soong should know better to take sleeping pills with alcohol. I'm worried that he is not responsive."

Raja thought for a moment and then bent over to poke David Huang on his chest. Before you can say Vladimir Putinsky, Raja was on his knees yelping in pain. David Huang had abruptly awakened from his stupor, and with lightning speed grabbed Raja's fingers and bent them backwards. Raja was now at David's mercy.

The pilot monitoring the passengers rushed out of the cockpit with a Beretta pistol in his hand, pointing it first at David and then at Victoria while shouting at the top of his lungs, "Let him go now or I'll shoot you both."

Victoria learned something about the man she was supposed to trust with her life. David was not the diplomatic or negotiating type. He had used Raja's body to shield his loaded Glock. Before the pilot finished his sentence, he was already dead. David fired three quick shots at the man's chest, just enough to stop him on his tracks. Besides being a master of disguise, David was also a skilled killer.

It took a while for Victoria to recover from the shock of seeing a comatose David spring to life, execute a lightning fast finger-hold, then hit three bullseyes on a moving target with his pistol. The entire sequence took less than a second.

David let go of Raja's fingers, but kept his Glock pointed at his erstwhile colleague.

"Why, my friend?" asked David, "What could they possibly have offered you to betray us?"

"They threatened to harm my family," said Raja with fear in his eyes. "I have no choice."

"You should've known better," said David. "The only way to help your family was to tell me."

"I'm very sorry," said Raja. "Please give me another chance. You need me to land the plane."

"Raja my friend, this will be the last time we work together," said David. "Take the plane to the nearest airport. After we land, we'll go our separate ways, and I don't ever want to see your face again. Do you understand?"

"Yes, thank you, thank you," said Raja. He got up and walked tentatively towards the cockpit. David covered Victoria's eyes with one hand and fired two shots at Raja's back. Without making even a whimper, Raja slumped to the floor.

"Why did you do that for?" cried Victoria. "Didn't we need him to land the plane?"

"Don't judge me. Raja gave your whereabouts away causing your parents' death," said David. "He was also trying to notify his new masters of his failure to deliver us. I really don't have a choice."

David pried open Raja's fist to retrieve a small device, which he immediately crushed. "Now, we'll have to resort to Plan B."

"Sorry to be demanding, but whatever plan you're using," requested Victoria, "please give me a little warning the next time you're about to give me a heart attack. Please."

"Poor Raja didn't know how devious these people were," said David. "They contacted me through a trusted secret channel at the same time they were trying to compromise hm. It was cheaper and less of a hassle for them to pay me directly. I knew about Raja getting the Fentanyl, which is why I'm well prepared."

"But what about me?" asked Victoria, "Didn't he put that in my drink too? How come I don't feel sleepy at all?"

"Either he hasn't tried to drug you or maybe you're immune to it," said David. "Sorry about the violence. You may have already guessed that your uneventful and sheltered Canadian life is over. Your odyssey to the truth is not going to be a garden party."

David got into the cockpit and pressed some buttons on the instrument panel. Victoria could feel the plane descending. David then unzipped his large travel bag. Plan B was escape by parachute. He helped Victoria put on a harness. They would do a tandem skydive in the middle of the night and in the middle of nowhere.

"Well, chalk up another first on your list—tandem skydiving at night without a legal drop zone," said David, as he handed Victoria a helmet and a pair of goggles. "Here, put these on. Don't worry about the jump. I promise it'll be fun and safe."

David strapped on the parachute and then headed towards the baggage door at the back of the plane. He beckoned Victoria to follow. "Come on, we're approaching three thousand metres (nine thousand feet). It's time."

Victoria was very concerned about jumping out of an airplane. She had to trust her life to a stranger whom she had met less than 24 hours ago, who also happened to be a cold-blooded killer. But since that car crash on the highway, Victoria had learned to build inner strength by staying calm in the face of insurmountable odds working against her favour. What's the point of panicking when you're strapped to your seat at the pinnacle of the rollercoaster, about to plunge into the abyss at a thousand klicks per second?

David hooked Victoria's harness to his tandem parachute. Then he gave her last minute instructions, repeatedly assuring her it would be fun while assuaging her fears.

"Be ready for the wind. I'm going to open the baggage door," David warned as he pulled the latch over. The door slid open letting in a rush of cold air. The wind beating on Victoria's face reminded her of the freezing rain that woke her after the car crash. She shuddered at the thought.

It was pitch black outside. David was wearing a powerful headlamp, which he immediately switched on. He also had several glow sticks and a flashlight in his vest. "This is it," David shouted in the buffeting wind, and with a leap, the two tumbled out of the plane into the deep dark void.

CHAPTER 12

Running on Empty

Victoria and David landed safely beside a quiet roadway about five hundred metres from a solitary gas station with a convenient store. The rest was pure darkness except for the headlights of a single car so far away it resembled a faint star from another galaxy. The drop was, as David had promised, an unforgettable joyride for Victoria. The freefall, the icy wind, the rush of adrenaline, the feeling of not being in control and yet believing that all would be fine because a guardian angel was behind her, gave Victoria the thrill of a lifetime.

After disposing the parachute, they took a brisk walk to the gas station. The place looked desolate and deserted. A half-broken sign at the pumps said, #1 Corn… Gas. Some letters were missing after Corn.

Only one person worked in the whole joint. The name tag on his uniform said Brent. He was reading the comics and didn't enjoy doing business just before closing.

David and Victoria sat down after ordering a coffee and a hot chocolate.

"I have good news," whispered David to Victoria. "A car is coming to pick us up."

"How is that possible?" Victoria cried, "It's the middle of the night and we're in the middle of nowhere. I didn't see you calling anyone."

"You noticed that little dot of light on the highway?" said David, "That's our car. I'll explain later."

Not long after the drinks arrived, a black car screeched to a stop outside. It was no ordinary automobile but a futuristic space age vehicle, dubbed the F011. It looked like something that existed only in the imagination of alien engineers.

"Our ride's here," David said as he got up, left a hundred dollar bill on the table, and signaled Victoria to leave.

"Keep the change, Brent," said David to the surprised but pleased attendant. "We haven't been here, we're not here, and you haven't seen us."

Victoria admired and inspected the space-age vehicle from up-close. She was astonished, even gobsmacked, when the door opened by itself and she found no driver inside.

"She's a beauty, isn't she?" said David. "This baby is designed and built by some of the cleverest people in the world. Autopilot is only her most basic function. There is a lot more than meets the eye. The car's Artificial Intelligence has guessed with great accuracy where we would end up and has been waiting in the neighbourhood. Once I activated Plan B from the plane, she has been homing in on us."

David took off the GPS/Altimeter watch from his wrist, gave it a determined stomp underfoot, and then threw the remains to the side of the road. David had an expensive antique Breguet tourbillon watch, which he used for checking the time.

"Won't this car be too conspicuous?"

"Actually no," said David. "The car's exterior is made of a dynamic metamaterial that creates excellent cloaking effects when we apply the proper electromagnetic fields. It also has a camouflage function to help us blend in with everyone.

"Besides, by jumping out of the plane, we have shaken off the bad guys. Going by land now will really throw them off because they don't expect it. This car is immune to hacking, which is what these people would try to do. Our navigation system uses China's quantum satellite encryption technology that makes us invisible to any electronic trackers and impervious to even the most capable hackers."

"I don't mean to be a naysayer," said Victoria. "Last time I read the news, China's quantum satellite technology won't be ready until decades down the road. Are you pulling my leg?"

"Don't believe everything you read in the news," David said. "This technology has been around for some time. They just needed to have their Beidou (北斗; northern ladle, Ursa Major) GPS satellites and Micius (墨子: mo-zi; ancient philosopher) quantum satellites in place. Their system is up and our quantum channel is working fine."

As the seat belt automatically strapped down David and Victoria, the hologram of a life-sized woman appeared in the empty seat beside them. She was an attractive Eurasian woman in her late thirties or early forties. She had reddish hair and stunning green eyes.

"David, how are you?" said the electronic assistant. "Your Highness, I am Moira. Welcome aboard."

The hologram looked so real that Victoria tried to touch it to make sure it's only an optical illusion.

"Good to see you, Moira. Input destination coordinates for Vancouver Pan Pacific Hotel," said David. He then commanded the car's control system as if he was the captain of a spaceship. "Initiate D30 kinetic impact resistance and EM impulse shield. Activate electronic stealth, adaptive noise cancellation, optical cloaking, and daylight camouflage—make it Buick Lacrosse. Set cruising speed at maximum warp, and … engage."

"I'm impressed, Captain Picard," said Victoria, who was a Trekkie and binged on *Star Trek: The Next Generation* on Netflix.

"What is the propulsion system of the *NCC 1941*? That's the licence plate number if I'm not mistaken."

"Very observant," said David. "Everything in the Starship Bozeman is cutting edge technology. We have a hybrid solar power and hydro fuel system making use of a nano-metamaterial electro-catalyst and a graphene supercapacitor battery. In other words, we take power from the sun to break up water into hydrogen and oxygen for emission-less fuel. We also have a parallel power system that uses a Lamarr full-spectrum phase-array gractenna to convert ubiquitous gravitation and electromagnetic energy into electricity using a levitated gyro-super-magnet with a shaped field. This concept model is officially known as F011, custom designed and built for me by secret associates at Daimler. This is the future. It runs on empty … forever."

CHAPTER 13

Warrior Queen

It was late in the day. The heavy skirmish yielded the desired result for Queen Zia. The rearguards of the northern barbarians known as the Gotts (鬼方: gui-fang; land of spirits) were finally driven off the hillock. The quick march of the queen's army surprised the Gotts. The heavily armoured Shang hoplites (foot soldiers) had the reputation of moving only slightly faster than molasses. The Shang people built caravansaries on their roads at about 7.5 kilometres (4.5 miles) apart representing a day's journey; that distance was known as a *she* (舍: lodging).

The Gotts normally moved at least three times faster, or more if they needed to, but on this occasion, they were bogged down by the large number of men in their horde and by their tremendous plunder. What the Gotts did not know was that the Shang king Wotan (武丁; wu-ding) with the bulk of the Shang army was at least fifteen days' march away. Wotan's wife Queen Zia (妇好; fu-hao), following the advice of her trusted counsel and field marshal Lady Diane, had decided to give chase with her smaller but more mobile personal army.

Although the hillock was just a grassy knoll in the middle of an otherwise flat landscape, it did command the view of the whole area all the way to the muddy river about five kilometres (three miles) to the north. It was also perfect for stationing reserve troops hidden from view of the enemy, so that they could move to the left or right to reinforce either wing without being seen.

The northern intruders were now facing an unenviable dilemma. They needed to return to their home base in the north, but trying to cross the river with all their plunder while Queen Zia's army was hot at their heels invited disaster. They could give up the loot and disperse, but they had come a long way and no one wanted to go home empty-handed.

The river bent to the south towards Queen Zia's right wing, forming a natural defence for the horde's left flank. The only escape route for the Gotts was to the northwest, where the river bent northwards. But a village stronghold and thick woods stood in their way. In any case, marching across the front with their left flank exposed to attack by Queen Zia's army was too risky. It seemed the Gotts had no choice but to stand and fight with the river behind them. They preferred to fight anyways. Death would be better than running home like dogs with nothing to show for after raiding deep inside Shang territory.

While the dreaded Gotts had pillaged and razed Shang villages with impunity, they did that without facing any serious opposition. Now that Queen Zia's army had arrived, they must contend with well-trained and armoured hoplites advancing in impenetrable phalanxes (formations), wielding long Shang halberds (戈: ge; OB: ) made with shiny bronze. They would encounter the queen's crack troops, which were famous for slashing through enemy ranks with their fearsome gladius bronze swords. They would have to brave the shower of arrows released by the queen's archers, who used powerful composite bows that could shoot arrows through armour at two hundred metres.

The Gotts with their legendary whirlwind cavalry charges were accustomed to running rough shod over Shang village defenders. The queen's army would be another story. War-hardened aristocratic commanders in four-horse chariots led her well-trained and disciplined hoplites. The young queen's soldiers would be as immovable as a mountain against the screaming and howling Gotts in their fearsome wolf masks. The Shang soldiers had reason to feel confident. They had not suffered any battlefield defeat under the leadership of Queen Zia and her handmaid, the resourceful and indomitable Lady Diane.

Queen Zia and Lady Diane arrived at the top of the grassy knoll, each in a royal four-horse chariot. The queen's horses were brilliant white steeds. Lady Diane's horses were shiny black. A phalanx unit of the Royal Guard numbered at seventy-two hoplites accompanied the two female commanders. Of the Royal Guards, twenty-four female personal guards provided the first ring of defence; the rest were special eunuch soldiers who had their left eye pierced and then trained to be superb marksmen with their composite bows.

The Shang word for 'slave' (臧: zang; OB: ) is a halberd stabbing an eye, while the word for 'household-slave' (童: tong; OB: ) is a man having an eye blinded by a dagger. Both words have adopted new meanings in modern Chinese. Zang (臧) now means good, and tong (童) means kid.

In addition to the charioteer, each royal chariot had an attendant known as the lictor. She was a trusted servant and a personal guard who carried the queen's fasces (钺: yue)—a large shining bronze axe on a pole weighing a total of about twenty-six to thirty pounds, with the *Lord of Animals* design and the queen's royal name cast on its surface. The Chinese character for minor king, duke, or lord (王: wang; OB: ) came from the shape of the Shang fasces.

The fasces was the symbol of Queen Zia's aristocratic and military authority. She had the sole discretion to take the life of a person under her rule. Her subjects lived at her pleasure. On the other hand, the queen was totally responsible for her subjects' well-being. With great power comes great responsibility.

Lady Diane was Queen Zia's most trusted officer and closest companion. Her chariot was a gift from the queen. She also had a lictor holding the queen's minor fasces. It was a slightly smaller axe signifying Lady Diane's subordinate position. Lady Diane had her own fasces, which she used while reigning in her own dukedom.

Queen Zia was an extraordinary Shang royalty. Not only was she the highest ranking wife of the Shang king Wotan, who maintained a harem of over sixty wives and concubines, most of them the result of alliances, she was also the Royal High Priestess, as well as the king's best general. In an earlier war against the western barbarians, King Wotan entrusted her with the largest army ever assembled by the Shang Kingdom. Queen Zia had an advantage over other wives; she was King Wotan's blood kin.

Queen Zia was a fighting general. She wore a full-body armour made of bronze plates and donned a shining bronze helmet with two tall pheasant feathers waving high in the air. Lady Diane was similarly attired but her helmet had smaller eagle feathers. The queen wore a white cape while her second-in-command wore black. Both Queen Zia and Lady Diane were beautiful and nubile girls in their early twenties. In another day and age, they could be modelling haute couture on runways in Paris, Milan, and Shanghai. Although considered young by modern standards, Queen Zia had already given birth to two princes and two princesses, guaranteeing her status as the mother of the future successor to the Shang throne.

"What do you think?" Queen Zia asked Lady Diane. "Will tomorrow be a good day for battle?"

"It'll be the best day ever," Lady Diane replied. "The gods have spoken. They're on our side."

"Will trickery in divination always bring us victory?" asked Queen Zia with an uneasy smile. Both Queen Zia and Lady Diane were official augurs and knew the tricks of the trade. "If we wait for the King, we can be more assured of the outcome."

"Have faith, Your Majesty," said Lady Diane. "Even if you have doubts about the gods, you can have faith in me and your army. It is not every day we can trap the elusive Gotts. Let us prepare for battle tonight. We will have a great victory tomorrow."

Lady Diane's network of scouts and spies was her secret weapon. Based on intelligence reports, she made maneuvers that forced the Gotts to fight a decisive battle at a location where the barbarians least wanted it. Lady Diane understood why Queen Zia was not as enthusiastic as she was. After all, the young queen was a long way from home, chasing after a fierce barbaric tribe at the edge of the kingdom. Besides, Queen Zia was a mother; she must be missing her babies. Wars and battles were always unpredictable. King Wotan was a great warrior and yet gained no advantage in past encounters with the wily Gotts.

"Come to my tent and have a drink with me," said Queen Zia. Like all Shang, the queen was a big drinker. Her wagon train of wine amphorae was legendary. "I brought your favourite millet beer and honey wine. I also want to give you a good luck charm for the battle. You'll sleep over at my tent tonight. Tomorrow we fight."

"Yes, Your Majesty," said Lady Diane, "but I would like to stay here for just a bit longer. I'll join you after sundown."

"Don't be long," said the queen, as she and her retinue retraced their route down the hillock.

Lady Diane made a gesture; her personal attendant quickly brought over a stepping-stone called *deng* (登: ascend; OB: ) for her to get off the chariot. This *deng* stepping-stone would someday become the family name of Deng Xiaoping, who would use the stepping-stone to bring China across the river.

The sun was sinking rapidly in the west. The Shang soldiers were saying their evening prayers and making their offerings to the setting sun. The Shang believed in many gods, and the sun was one of the most important. The name of Shang dynasty's founding king was Tang (汤; OB: ), which is the symbol of the sun and its benevolent rays.

Ancient myth tells of a time when the heavens had ten suns and that on each day, only one of them came out in turn. One day, all ten suns came out, stopped in the middle of the sky and refused to leave, scorching the earth and torturing humanity, until the hero Houyi (后羿), husband of the moon goddess Chang'e (嫦娥), shot down nine suns with his divine bow and arrows. Since then, the one remaining sun has been rising and setting every day in an orderly fashion.

A scholar of antiquities by the name of Immanuel Velikovsky, famous for writing the book *Worlds in Collision*, thought that the myth recorded a time when the earth stopped spinning, and the sun did not set for ten days, causing worldwide catastrophes. It is a mystery how the ancients knew it was ten days if the sun didn't set. They didn't have Breguet tourbillon watches at the time.

Lady Diane studied the enemy camps and noticed how they were preparing for battle. She had heard that the Gotts were undisciplined wild men on horseback. Lady Diane saw something different with her eagle eyes. The Gotts were well armed, well organized, and well prepared. But it made no difference because she would defeat them just the same. Lady Diane was a master in the art of war, the kind that was unknown to people of that era.

She looked to the south to enjoy the picturesque view of the great kingdom of Shang. Dozens of villages in the distance were sending columns of smoke up into the heavens. The Gotts spared these lucky villages because they had to run from the queen's forced march. This was mid-autumn; the leaves were changing colours and harvest was under way. The smoke came from the burning of the stalks of harvested crops. It would repel locusts and birds. The ashes would also become fertilizer for next year's crops. The Shang word for autumn shows a locust and a fire (OB: ), later adding a sheath of wheat. The modern word for autumn (秋: qiu) omits the locust, leaving a composite word made up of wheat and fire.

It was Lady Diane who first convinced Queen Zia to volunteer for the expedition against the Sand People (土方; tu-fang; land of earth) in the north. At that time, Queen Zia was only one of King Wotan's more than sixty wives and Lady Diane was her handmaid. Both girls would likely waste their lives away in the harem.

The opportunity came when King Wotan had a tough time dealing with the Sand People. After receiving a number of entreaties from Queen Zia, King Wotan finally relented and gave her fifteen hundred of his best men with fifty chariots to lead into battle against the Sand People. Queen Zia's father, Duke Yarjee (亚其: ya-qi), cousin of King Wotan, would lend her a thousand soldiers and twenty chariots from his fief. King Wotan however did not have high expectations that his wife could succeed whereas he had failed several times with larger armies.

When the news reached the capital that Queen Zia's army crushed the Sand People, razed their capital, captured the ruling clan, enslaved a thousand young men in the flower of their youth, and was returning home laden with loot, King Wotan travelled a distance of three *shes* (舍: lodging) to welcome his favourite wife back home. After that, Queen Zia would monopolize the dragon's lair and bear the king two sons and two daughters.

According to the traditions of war, King Wotan granted Queen Zia large tracts of fertile land and industrious subjects where she reigned as an independent vassal. Queen Zia however did not reside in her own realm but sent personal servants to administer it.

The stars began to dominate the evening sky as the last light of the setting sun faded away. Lady Diane gazed longingly at the prominent constellation of the giant hunter Orion. She felt a wave of sadness as hidden memories emerged from the dark corners of her heart. She remembered waking up in the wilderness wrapped in a black shroud, feeling the heat of a raging fire underneath her skin. She did not know where she was or how she got there. It was her good fortune that Queen Zia found her, healed her, and kept her as a royal handmaid.

The queen later explained to Lady Diane why she decided to save the life of a dying slave. She saw a red tattoo marking on the dying girl's forehead. The Shang word for 'concubine' (妾: qie), which meant female slave in Shang times was the drawing of a tattoo knife on a woman's head (OB: ). Red tattoos however had a special meaning for the Shang. They were reserved for the dead. The Shang honoured their deceased with the title *Wen* (文) which meant tattoo. Lady Diane must have overcome death in an ordeal. A person who defied death was meant to achieve great deeds.

Since Queen Zia's status rose to stratospheric heights, Lady Diane's fortunes improved as well. After all, she was the queen's chief strategist, top military officer, and field marshal in all of the queen's victorious battles. When the queen received her own vassal state, Lady Diane also received from the king her share of choice fields. The queen named her Diane, which came from the aristocratic title *Dian* (甸) in relation to the choice fields she was getting near the capital. In addition, the original character for *Dian* was sometimes indistinguishable from the character for 'field' (田: tian), which also meant hunt. Indeed, Queen Zia found the mysterious girl in the middle of a royal hunt.

In those days, it was highly unusual for an unmarried woman to receive a title and to own property, but Lady Diane was an unusual person. Furthermore, Queen Zia had granted Lady Diane a four-horse royal chariot and a lictor with the queen's fasces to boot. It would be wise for men who had some wrongheaded belief in women's place in society to keep their opinions to themselves and their mouths shut, lest they ventured to feel the wrath of the queen's fasces on their necks.

Queen Zia had wanted to arrange an appropriate husband for Lady Diane, but she could not persuade her pretty handmaid to marry. Lady Diane's secret was that she already had a family, but she might never see them again. At times, she needed to be alone so she could secretly mourn her loss. Lady Diane was no nameless servant; neither would she ever be a concubine to some Shang lord. She was a true blue royalty trained since birth to be a superior warrior. Lady Diane took off her helmet to reveal the red tattoo at the centre of her forehead. It was the same heart tattoo on Victoria's back.

CHAPTER 14

The Dialectic

Victoria didn't know what to make of David and his role in turning her life upside down. She intended to find out. "Mr. Huang, I want to thank you for risking your life for me," said Victoria. "But now that we're in outer space with no prying ears, will you please answer my question? I'm dying of curiosity."

"It's my honour to be of service, Your Highness," David said. "As for the rather incredible story of who you are, how we end up here, and where we're going, I'll have to find some solid ground on which to begin."

David paused for a few seconds in silent contemplation. "I suppose," he said, "it must begin with having you first accept that everything you have known so far is a lie."

"I'm sorry but I don't agree. I've heard that cliché before," said Victoria. "If as you say everything I know is a lie, then what does that make you?"

"You're right, lies can be insidious," David said. "In real life, fraud, deceit, and malice are usually mingled with truth and sincerity. You should be careful. I suggest we play a game of dialectics."

"What game is that?"

"We decide on a subject," said David, "and then try to discover a mutually acceptable truth through discussion. Socrates and Hegel used to play the game. Are you game?"

Victoria loved playing games; she was naturally good at games. Michael taught her to play chess when she was just starting first grade. By the end of the week, she was already employing multiple variations of Sicilian Defence and complex Nimzowitschian hypermodern openings, sometimes making breathtaking king side pawn attacks, and declaring checkmates in twelve.

Angela taught Victoria to play backgammon. After learning the rules, Victoria went on to beat her mom eleven games in a row … no mercy. Michael and Angela had some idea about Victoria's natural abilities, which was why they did not allow her to play skill-testing games outside of home. The word 'game' however lit up Victoria's neural networks.

"If it's a game, then I'm game."

"First topic today: Does everyone lie?" David said. "Let me start with a few questions. Who do you trust the most?"

"My parents, naturally."

"Do you think they have ever lied to you?" asked David.

"I suppose they have lied by omission," said Victoria after thinking for a few seconds. "They never told me about my background, but I believe they did that to protect me."

"Good, we agree that the people we trust most will lie to us," said David. "Next question. In your opinion, which is the most important book in the world?"

"It's a no brainer; the Bible of course."

"What does God of the Bible say about lying?"

"Thou shalt not bear false witness against thy neighbour."

"Nevertheless, the Lord thy God told lies," said David. "In Genesis Chapter 2, God told Adam that '… of the tree of the knowledge of good and evil, thou shalt not eat of it: for in the day that thou eatest thereof thou shalt surely die.' The serpent on the other hand told them '… ye shall not surely die … your eyes shall be opened, and ye shall be as gods, knowing good and evil.' We know what happened. Adam and Eve ate the forbidden fruit, they did not die on the day they ate the fruit, and their eyes were opened. Would you agree that God lied?"

"But God was testing Adam," Victoria said. "And He was talking about the death of the sinner's soul, not his body."

"We're not discussing the morality of lying or religious interpretations. I just want to confirm that even God lies," said David. "Let us move on to the next question. Who is your favourite U.S. president?"

"Let's see, I think I'll skip the Republican presidents. I'm too young to be conservative," said Victoria. "Obama gave a lot of hope with 'Yes We Can,' but that turned into disappointments. Even Michael Moore eventually said how useless it was to have hope but no action. Obama also didn't do anything to earn that pre-emptive Nobel Piss Prize either—so, not Obama.

"Bill Clinton bombed Yugoslavia, the Chinese embassy, and a medicine factory, among other distasteful acts—I'll pass. Jimmy Carter; another Nobel winner, is better as an ex-president. Lyndon Johnson has Vietnam War and the Tonkin Gulf Incident to his name—no thank you. John F. Kennedy was another belligerent president who brought us to the brink of Armageddon. He also badly underestimated the enemies of peace, paying for the mistake with his life. Harry Truman dropped atomic bombs, covered up Camp 731's heinous crimes in China, waged germ warfare, and gave us the Cold War. With liberals like that, who needs conservatives? I guess that leaves us Franklin Delano Roosevelt. He's the one for me."

"So, you know your presidents. But I would like to add that a large portion of FDR's family wealth came from his maternal grandfather Warren Delano, Jr.," David said. "During the second half of the 19th century, Warren Delano was the CEO of the largest American trading company in China. He made a ton of money selling opium among other less odious merchandise. FDR's path to success was paved in part with drug money from China."

"That maybe so," said Victoria, "but that's not FDR's fault."

"Not finding fault at all," said David, "just stating a fact. There were other big families doing it, such as the Russell, the Perkins, and the Forbes. The money also did a lot of good, building American railways and contributing to centres of higher learning such as Harvard, Yale, and Princeton. Delano drug money supported the rise of one of the most liberal presidents in the history of America. Was it somebody's fault that China during the Qing dynasty could not defend its own rights? The English sent only fifteen thousand soldiers into China during the Opium Wars and got the Qing court to completely capitulate. The Americans sent half a million into Vietnam and went home with an experience they'd rather forget, and they probably have. Of course, Vietnam got a ton of help from its neighbours and allies.

"After the Opium War, Chinese intellectuals saw the need for change if the Chinese civilization was to survive the onslaught of the West. Come to think of it, opium was what woke the Chinese dragon from its stupor."

Looting of the Old Summer Palace (1860)

CHAPTER 15

Dante's Gate

Victoria could not understand why David was taking her on a random walk down the garden path. How would biblical criticism and historical revisionism help her understand her background? When she became impatient and tried to take a shortcut to the final destination, David steered her back each time, keeping her on the straight and narrow, explaining that she must travel the entire length of the odyssey.

Whether it is for Dante going from the Inferno through Purgatory to Paradise, or the Little Prince travelling from asteroid B-612 to earth, or Gary King wending through the Golden Mile to the World's End, the path travelled is more important than the destination. Our minds have so inured us to the lies that we will not be able to accept the truth, unless we first cure our minds of the sickness by walking down the garden path.

David went on to clarify a few more details about FDR. For example, contrary to the cherished belief perpetrated by Western historians and Hollywood movies that the U.S. and its Western allies defeated Nazi Germany and saved humanity, a lot of the credit should in fact go to the Soviet Union.

By the time American soldiers and their Western allies landed in Normandy, the game was almost over for Hitler. The German side had lost about one million irreplaceable soldiers after the surrender of their Sixth Army to the Soviet Union at Stalingrad in early 1943. Then at the Battle of Kursk, Hitler suffered another irreversible defeat to Stalin, leaving Germany with little resistance against the Soviet advance into Berlin. All these had come to pass by the fall of 1943, about nine months before the Normandy landings. While waiting for the second front promised by its Western allies to materialize since 1941, the Soviets crushed the German war machine on the eastern front with little direct outside assistance.

Furthermore, after the surrender of Germany, the Soviet Union entered the Pacific theatre, and within ten days of starting the campaign, destroyed the Japanese Kwantung army in Manchuria, which numbered over 700,000 soldiers supported by 200,000 local recruits, leaving Japan with no alternative but to surrender unconditionally.

Back in America, FDR made a compromise that took the U.S. down the slippery slope. At the 1944 Democratic Party Convention, FDR acceded to the Democratic Party barons' pernicious plan to install Harry Truman as VP, stealing the nomination from the vastly more popular Henry Wallace. A group of powerful elites who called themselves democratic applied Machiavellian machinations to thwart the will of the people. The world would have been a different place if Wallace had become President when FDR died in office. Wallace, after all, was a peaceful peasant who refused to make backroom deals to gain power.

Wallace might have regretted his naivety in later life. If he had not trusted so strongly in the principles of democracy and fair play, he might have had a chance to make the world a better place. Imagine a world history without the Cold War, or Korean War, or Vietnam War, or all the other endless senseless wars with no names and against shapeless enemies. Imagine a world based on peaceful cooperation, mutual respect, and sharing, even with the weak, the dispossessed, and the underprivileged.

Poor Wallace should have noticed that he did not have democracy in his own home or his own company. How likely it was for powerful barons to cede power to someone they detested? The manipulation of the people's will for political gain however started democracy's voyage on the river of no return. It is inevitable. Aristotle said so. The ancient Greeks had a lot of experience.

"Sometimes when we find ourselves trapped and there's no way out," David said, "it's because we have passed through the infernal gates ignoring these words: Lasciate ogni speranza, voi ch'entrate—Abandon every hope, all those who enter."

"There is always hope," said Victoria, who knew the origin of the quote because she was a voracious reader and had consumed a hundred classics by the time she was ten. "Dante went through those gates and survived well enough to write a bestselling trilogy about the trip."

"But he had the help of Virgil the poet and Beatrice the pious girl he worshipped," David said. "Who'll help us?"

"You're helping me," said Victoria, but then she wasn't quite sure. "Aren't you?"

"In truth, we'll have to help ourselves," said David. "I'm doing a bad job if you trust me blindly. We must not be fooled by liars and their lies. We should not take other's words for granted, especially from those we have been taught to trust. When people and institutes lie to us all the times, they will not suddenly decide to tell us the truth unless it is to confuse us. We should also examine our own prejudices. I supposed we can't learn the absolute truth and we can't avoid lies, but we should at least be aware of them and be diligent in learning the truth."

"One of my favourite writers, I. F. Stone," said Victoria, "gave similar advice."

"I. F. Stone was the one who opened my eyes," said David. "I now read regularly from over a hundred different writers. They include Seymour Hersh, Glen Greenwald, Uri Avnery (RIP), John Pilger, Pepe Escobar, Chris Hedges, Jay Janson, Robert Reich, and Matt Taibbi just to name a few. I also read from the other side of the spectrum like Ron Paul, Ron Unz, Robert Parry (RIP), Lew Rockwell, Justin Raimondo, Pat Buchanan, Paul Craig Roberts, Tom Engelhardt, David Stockman, and David Goldman. I read very mainstream writers such as Paul Krugman and Thomas Friedman, as well as those who hold strong critical views of current affairs in America, such as Gerald Celente, Noam Chomsky, and Andre Vltchek. I make sure I get an idea what everyone is saying. Learning the truth is hard work."

Dante's Inferno: Titans and giants
by Gustave Doré

CHAPTER 16

Paradise of Shangala

Armed only with a walking staff and draped in his cloak of black feathers, Chief Deskull walked into the wall of fire that separated his army from his archenemy. By his side were his two unicorns Galizu and Faligu, meaning 'peripatetic' and 'inexorable' in the local language. Chief Deskull was the warlord of the Legion de Lude. He was the leader of an army made up of the Animorta—the Soul-Dead. Most of his soldiers were victims of the Endless War that had torn their body and soul apart. Chief Deskull wore a skull helmet that covered his left eye. In place of his eye was a *Universal Spectral Scanner* (USS), which gave him a profound perception of his surroundings, an ability his good right eye did not possess.

When Deskull became the chief of the legion, he received the mantle of leadership known as the Kanaka, a black cloak covered with the black feathers of the Equilas, a mythical bird born of water and fire. He also received the mysterious Nephrusta, the staff of power. Both relics were passed down from the ancient sage Santos Didymas, the founder of the order.

Myth said that Didymas emerged from a rock. People of his time called him the Stone man. He travelled alone in a chariot drawn by a unicorn named Lude, meaning 'one-horn.' The speed of his chariot was legendary. He could travel from one end of the planet to another in a matter of seconds. The sage Didymas went from one battlefield to another saving the wounded soldiers and repairing their bodies and their souls. From them he created the army of the Animorta.

Chief Deskull's twin sons Sheru and Sachem marched behind him to the left and the right as his wing commanders. Both boys were still in their teens but had already experienced enough killing and bloodshed of several lifetimes. The name Sheru meant *Covenant of God*, while Sachem meant *Stygian Oath*. Chief Deskull had sworn oaths of heaven and hell that he would avenge the death of his parents and his sister. His sons' names served as constant reminders.

The wall of fire was extremely hot and could deter the passage of any living organisms. It was named the Eternal Inferno because the technology to extinguish the flames did not yet exist. It also created an impenetrable electromagnetic barrier. No wall however could stop Chief Deskull and his army of robots, androids, automated warbots, bionic anthropods, and soldiers clad in exoskeletal battle gear. Chief Deskull knew that waiting on the other side was the invincible Yarjun, his last obstacle to global rule and the fulfilment of his oaths.

The Eternal Inferno was created by a fuel mined from the core of this bountiful planet known as Shangala. This highly flammable material was called Erinysite, which meant the 'flames of fury.' It was the cheap and abundant starter fuel for the planet's fusion reactors. It also gave the flames multiple colours ranging from purple to blue, to green, and even to orange and yellow.

Erinysite was a blessing because it made nuclear fusion possible and economical, and thus gave the planet's inhabitants a virtually inexhaustible source of clean energy, from which was produced an unlimited supply of food. Economics for life forms became moot when the supply of energy and food was unlimited. It was also a curse because it led to unending war.

Wars usually stop when economic conditions prevent their continuation. Wars and the economy that supports them ultimately depend on food. Technological development finally removed that constrain. Wars on Shangala henceforth became endless, eternal, and existential.

The final battle must be fought. Many years ago, the Yarjun had saved Chief Deskull from certain death while he was still a young lad, gave him whatever his heart desired, and taught him everything he knew. The Yarjun considered Chief Deskull a member of the family, and yet he betrayed the Yarjun when he grew up. Since then, the Yarjun had become an inveterate enemy of Chief Deskull.

The Yarjun were born warriors. They belonged to the highest caste of the social hierarchy on Shangala. Yarjun soldiers lived and died for war. They had the mad but ingenious Stranghate Option as their insurance against peace. Peace to the Yarjun was the same as defeat.

Shangala was the oldest planet in the galaxy. It received an abundant amount of benevolent rays from a red supergiant known as Yin-Di-Ra the Lifegiver. The orbit of Shangala was so far away from the sun that one Shangala month was equivalent to approximately twenty-eight of earth's years, and there were three months in one Shangala year. The average lifespan of an anthropod, the dominant life form, was a hundred and ten years, equivalent to about 9,240 earth years. No one on Shangala ever complained that life was too short. The conditions on the planet were not only supportive of life, but encouraged evolution. Life had arisen from the primordial soup, evolved into highly intelligent beings, and become extinct several times over at Shangala.

Eventually, the solar system that supported Shangala became completely stable. The ecological system of Shangala also achieved perfect equilibrium. The largest and the smallest creatures, the flora and the fauna, the predator and the prey, all became interconnected and intertwined in a highly harmonious nexus of peaceful coexistence.

Mother Nature moderated appetite and avarice according to the availability of sustenance. She vitiated the violence and vicissitudes of the unpredictable future. Even evolution slowed to a crawl. Nothing needed to be changed or improved when life was already perfect.

At the beginning of this Golden Age, Shangala was a land of exquisite beauty. Its rolling hills and grassy knolls were adorned with small trees laden with colourful juicy fruits known as the bassilix. The taste of the bassilix would change according to many factors, such as the mood of the person who was eating it and the temperament of the people close by. Every mouthful was a unique experience. Eating the fruit would neutralize the most deadly poison, rejuvenate the aged, and extend one's life span.

The grasses of the meadows were as colourful as the rainbow dragon Irin. Flourishing all over the land were the perennial bushes of iridescent blooms known as Diamaranth, the immortal flower of Shangala. When the flower buds opened during the unpredictable triple lunar alignments, soul-soothing fragrances of Diamaranth would fill the air. Whoever inhaled the aroma would shed all memories of hate and fall in love with the first person who came into sight. Fairylike butterflies with diaphanous wings danced and pirouetted from one bloom to another, drawing the nectar known in the local language as Ambergrace. Those who tasted even a drop of the elixir would enter a state of Elysian ecstasy.

Crystal clear springs sprouted everywhere, creating babbling brooks that meandered through the land sustaining life wherever they went. The water was sweet and refreshing, full of nutrients, and had healing properties. Those who drank the nourishing water would never get sick.

The only large shade tree of the planet was the Ailantus, also known as the *Wind-chime Tree of Paradise*. It was the home of the Seraphana songbirds. Whenever the warm breeze Flavonus caressed the leaves of the Ailantus, a harmonious symphony of heavenly harps would reverberate all over the land. The Seraphanas, which had the voice of angels, would sing the majestic chorus of Gratia Plena (Full of Grace) in the praise of the almighty God Diamaranth, who had made the paradise of Shangala possible.

Each evening, the three moons of Shangala would waltz into the night sky, reeling in unpredictable circles around the Milky Way with their bewildering dance, occasionally accompanied by meteor showers that resembled New Year's Eve fireworks.

If God was experimenting with complex life systems, Shangala was a resounding success. Why God would create so-called sentient beings who immediately declared themselves lord and owner of all, then tirelessly slaughtered each other, even among brothers, while they unceremoniously destroyed God's creations, sometimes in God's own name, is one of the greatest mysteries of the universe.

It happened on one uneventful day in Shangala, when an insignificant simian-like creature accidentally discovered the expediency of a strategically held rock. Nietzsche, Strauss, Kubrick, and Clarke must have laughed with the gods when they saw this from their heavenly thrones at Mount Olympus. The four had collaborated in the creation of the Sci-fi classic *2001: A Space Odyssey*.

With an abundance of food, the simians multiplied and evolved. With free time on their hands, they spent their hours solving problems with their extraordinary brain. They soon learned to improve the effectiveness of their primitive tool by fastening the rock to a stick. Then they figured out a way to sharpen the rock and turned it into a lethal weapon. They began to organize and hunt in groups. They learned to enjoy the taste of blood. They turned into predators with an insatiable appetite. When they wanted more food, they learned to farm. Then they organized large groups to build walls and strongholds. Cooperation necessitated communications and language. From language emerged writing, and from writing civilization. All the while, they learned to slaughter each other in prodigious numbers by ever more innovative means.

Each step forward created more power and more successes for the simians. These creatures soon lorded over all flora and fauna, and they gave their subjects names. They would declare ownership of all the land, the minerals, the waters, and the air, including all that lay dormant in them, all that lived in them, and all that grew out of them.

That insignificant simian had inadvertently disrupted the harmony of nature. The ripple would eventually grow into a ferocious storm that engulfed the planet. These simians scientifically known as Anthropodius proudly called themselves Haryan, the children of Diamaranth, the people of the immortal flower.

As Chief Deskull entered the wall of fire, his skin began to sparkle. Chief Deskull's bionic epidermis was resistant to the heat of the Erinysite blaze. His unicorns Galizu and Faligu frolicked in the flames as if they were fish in water. The unicorns of Shangala were born of time and fire. Their birthplace was the Stygian river of Erinys. Erinysite was mother's milk to the unicorns.

Having demonstrated that the flames of the Inferno could not harm him, Chief Deskull raised his staff of power and the flames promptly extinguished, generating a mountain of impenetrable smoke that masked the army behind him. The unicorns had the secret power to put out the Flames of Fury.

Chief Deskull's staff of power, the legendary Nephrusta, was made from priceless nephrus, which came from the root of the Deathrider, a mythical giant tree with roots that encompassed the entire planet. The Nephrusta was crowned with a globe fashioned from the hardest material in the universe known as Sandymium, a very rare material that so far had only been discovered in Shangala. Nephrusta was unique in its existence and extraordinary in its powers.

Nephrusta was a word with no parallel in Earth languages. It meant everything and nothing; everywhere and nowhere; infinitely large and infinitely small; infinite in the past and infinite in the future; the creator and the destroyer. No one understood it or knew what it was because there was no means to probe it. No one knew how the globe was fashioned, since no tool would be hard enough to work Sandymium. In their ignorance, people just gave it a name, and they called it Nephrusta, in the same way that some humans called phenomena that they couldn't understand dark, strange, spooky, ghost, God, or the Force.

The supreme Yarjun warlord Killiam, whose name meant *God's Beloved*, was waiting for his archenemy in his impregnable Fort Destruck. He had read reports about Chief Deskull's earlier conquests. He knew Chief Deskull wouldn't meet him in battle without a few tricks up his sleeve. Killiam's Yarjun soldiers were well prepared. Their powerful and fearsome weapons of massive destruction were loaded and cocked, ready to fire at the command of their exalted leader. They had already said their death prayers. Death in battle was what every Yarjun soldier prayed for. Death was what every Yarjun lived for.

As the smoke gradually cleared, Chief Deskull and his army came into view. Chief Deskull stopped and raised his majestic Nephrusta. At that moment, three large black birds swooped down from the sky, simultaneously making a loud squawk that reverberated all over the land of the Yarjun. Two of the birds landed on the horn of the unicorns. The third circled above Chief Deskull for a round or two before landing on the globe of the Nephrusta.

These were not real birds, because by this time, Shangala had already lost its entire flora and fauna save for some poisonous creatures of the dark, the two unicorns, and several sickly blooms of Diamaranth. Even the immortal bloom could not survive the Haryan culture of death. Chief Deskull's black birds were flying machines designed to resemble the Equilas, a winged predator similar to Earth's eagles. Chief Deskull called the two mechanical Equilas on the unicorn horns Hugin—thought, and Munin—memory. The one sitting on top of the Nephrusta was like the military standard of Imperial Roman Legions. Its name was Augin—sight.

Chief Deskull's Animorta soldiers had more than a hundred flags, representing all the tribes, clans, and warring states he had conquered. Chief Deskull used the emblem of Shangala as his own military standards in anticipation of ultimate victory. It was a stylized symbol of the Diamaranth bush.

One might expect Chief Deskull to bring enough soldiers to flatten the land of the Yarjun, but he brought only a thousand. On top of that, they were unarmed. Chief Deskull was famous for doing the unexpected, and Lord Killiam was expecting the unexpected, but for Chief Deskull and his army to walk into the jaws of the dragon unarmed was beyond the realm of stratagem. Either Chief Deskull was mad or he had a death wish.

While the Yarjun loved wanton slaughter, especially of defenceless civilians, including women, children, and their harmless electronic pets, they had no idea on how to slaughter an army of unarmed soldiers.

"Lord Killiam, as you can see, we are unarmed," thundered Chief Deskull. "We do not seek slaughter today. We seek parley."

"Traitor, ingrate, guttersnipe, maggot, lowest of the lowest, Vermin of Xylem, spawn of the Abyssum," Lord Killiam responded with a string of invectives. "The Yarjun delivered you from your early grave, took you from your lowest of the lowest musician caste, and elevated you to our highest warrior caste. We spared no expense to make you whole and to give you strength. We taught you the art of war and trained you to be a leader, treating you like our own son, and what did you do? You ran away to join the Animorta (the Yarjun soldiers to a man spat in contempt at the name). You led them to attack our vassals and protectorates. Now you lead your army into our land unarmed. What more is there to say? You seek death."

"Lord Killiam, I died when I was young, you cannot kill the dead," replied Chief Deskull in his thundering voice. "I am Animorta, the Soul-Dead, and the dead has no fear. I come today not to fight or to dig up the past. I come to make peace. I offer peace and I offer life for everyone here. I even offer forgiveness. Choose peace and we shall live as brothers and equals. Choose war and you will all perish. We will obliterate the Yarjun and expunge its name from memory. It will be as if you have never existed. You will deliver the offer to the Supremes. I will patiently await your answer."

The Supreme Council of the Yarjun was known as the Supremes. It normally made all the important decisions for the Yarjun and their dependencies. Since the Yarjun settled all conflicts by violence, council meetings sometimes became gladiatorial melees, causing even delays to straightforward rubber-stamping of orders for more wars.

"That's very generous of you!" said Lord Killiam, God's Beloved, "But the Supremes has made me dictator with full powers to conduct the war according to my sole discretion. There is no debate. You can have our decision now."

Chief Deskull knew full well how Lord Killiam would respond. Lord Killiam had a gun pointed at his archenemy's head. Why would anyone in that position make peace? Lord Killiam returned Chief Deskull's offer of an olive branch with a derisive remark.

"I have a counter-offer for you. I offer to send all of you to the Tartary Pit with our Integrated Independent Individually Identifiable Intelligent Infinity missiles. While we're talking, our 6IM system has targeted and locked onto each and every one of you. On my order, it will keep firing until there are no life signs and no energy fields coming from the targets. Any last words?"

Lord Killiam did not wait for Chief Deskull to answer as he stretched out his arms and looked to the heavens as if he was making a supplication to the gods for justice and retribution. That was Lord Killiam's dramatic signal for activating the kill switch of his intelligent weapon system. Thousands of smart missiles burst forth from the launchers, while the artillery and infantry of the Yarjun Defence Force also poured fire and fury from their Exponential Blunderbusses at Chief Deskull and his Animorta soldiers. The war was over in a few seconds. One person remained standing. He called himself King Kull the Benevolent, the one king to rule them all.

CHAPTER 17

Golden Lies

On the straight and narrow Trans-Canada Highway to Vancouver, David continued his assertion that 'everything is a lie.'

"Let us explore the lies in economics," said David, taking Victoria into muddy waters.

Economics is a subject which few understands, but on which everyone has an opinion, all of which will eventually prove to be wrong. World-renowned economist John Kenneth Galbraith, a son of the True North, once said, "The only function of economic forecasting is to make astrology look respectable." Even he had to tread carefully.

David Huang the art-and-antique dealer was no expert on the subject. On the other hand, the lies are piled thick in this realm and dire are the human consequences for ignorance in this subject. Victoria could no more avoid wading through this cesspool of lies than to continue believing that she was just a normal average girl who would live a simple and sheltered life in Toronto.

"Now if you're going to get into business," David asked, "and you want to produce something valuable which is always in great demand, what would you produce?"

"I suppose food," replied Victoria while rubbing her tummy. "I can feel a demand coming up."

"Moira already planned a stop for us," David said. "But getting back to the question of food being the most important product, if the law of supply and demand holds, why do you think that the peasants, the producers of this valuable commodity, have always been poor and down-trodden throughout history?"

"Sorry to sound impolite," said Victoria, "I think the peasants must not be very smart and they got ripped off."

"The truth isn't always sweet music," said David. "And I agree with you on this point. If the peasants are producing life-sustaining food but someone else is eating their lunch, they may have themselves to blame. You can imagine that without food, there is no civilization and the mightiest army will collapse."

"But what can peasants do?" Victoria exclaimed, "They know how to grow food but they don't know how to sell their products in the free market."

"The truth is, the free market is a myth; it does not exist," said David. "There may be a convincing illusion of it with everyone constantly repeating the lies. There is no free market when wealth is defined by money created by an authority, which decides how the money is created, distributed, and spent; none of that has ever been fair or free."

"What about gold and silver?" asked Victoria, "Even kings can't create precious metals out of thin air."

"Gold is better than paper promises," replied David. "But it takes a huge amount of effort, and therefore food, to discover the mines, possess the mines, extract the ores, and turn it into coins. Without food there is no money, any kind of money, and therefore, people who want money must first exploit the food producers. At the end of the day, be it gold or fiat, money is the prerogative of rulers. The peasants and the plebeians get to enjoy what trickles down at the pleasure of their lords.

"That said; money is necessary for the running of the complex modern society. Unfortunately, in the hands of some clever but selfish elites, it becomes a means for them to steal from everyone especially from the wealth producers and still call it a free market. The game has always been rigged by the powerful and the privileged few while being justified by the lie that it's fair, free, and open.

"It's like setting up a hundred metre dash for a well-nourished professionally trained athlete against a starving and crippled child. Furthermore, the athlete has his starting line moved up by fifty metres and he gets a five-second head start. For insurance, the athlete also pays all the judges. Since we believe in bald-faced lies about the free capitalist markets, we can also believe that this is a fair race."

In her mind, Victoria questioned David's assertion that the system was unfair.

If the capitalist system is unfair, why has it thrived for as long as there has been a human economy? If the elites are paying lip service to serving the people while they are ripping everyone off, then why haven't the people done something about it? Is it normal for people to accept extreme inequalities in society? Is it natural for the top 1% of the population to own 35% of the nation's wealth while the bottom 60% owns only 4%? Is it because the top 40% allows it to happen? Is it because the only narrative we hear is to aspire to become the 1%? There must be some Svengali magic that mesmerizes the people into believing that this imbalance can be and should be sustained indefinitely. Someday, people may wake up from the stupor and find that there is only air underneath their feet, like when Wile E. Coyote overshoots the precipice.

The answer is similar to the exploitation of the peasants. Extreme inequality exists because the people allow it to happen. Why else would the people willingly pay taxes and go into debt to fund endless wars and support atrocities, including the wanton slaughter of children without requiring any debate, fiscal responsibility, accountability, or oversight, while they refuse to fund education, healthcare, and social security?

Half a century ago, Reverend Martin Luther King, Jr. said, "A nation that continues year after year to spend more money on military defence than on programs of social uplift is approaching spiritual death." He also said, "A man can't ride your back unless it's bent." In the end, he held himself and all Americans responsible. Was he speaking to a wall?

According to David's theory of the omnipresent lies, people who live and die in bullshit cannot comprehend flowers. People who have known only lies do not know what they want. If they know what they want, they do not know how to go about making the changes. Even if they know how to make the changes, they do not have the resources and the courage to fight the elites, who control the money and the narrative, and who own enough henchmen to pulverize anyone stupid enough to move their cheese.

So what happens when the elites make life difficult for the 99%? The people would finally spring into action and look for easy targets to put the blame on. They blame the politicians; they blame the other party; they blame the government; they blame the president; they blame Goldman Sachs and the Fed; they blame Walmart and Amazon; they blame Google and Facebook; they blame Artificial Intelligence; they blame trade; they blame immigrants; they blame communism; and of course, they blame China.

But the people never blame themselves for electing the same elitist politicians year after year, believing in the same lies over and over again, hoping the next demagogue with a silver tongue and a clever slogan will make everything great again, then trusting them with the power to squander the wealth and the goodwill of the people with impunity. What result should one expect?

"But how do people get this elusive change," asked Victoria, "that they've been looking for?"

For a start, people should just stop lying to themselves and start admitting that they are themselves to blame. It is a government of the people after all and the people are responsible. Blaming others and going to war will not help. It will only hasten self-destruction. If there is a problem with the system, then it is up to the people to fix it, not trying their best to export it. If they cannot fix the system, then the people must come up with a new paradigm. They should stop believing in the lies and start thinking deeply and creatively.

Unfortunately, the people are mesmerized by the so-called free press that is mainly a federation of feckless factories of fabrication that fans the flames of violence. Upton Sinclair said this about the journalists of the free press, "It is difficult to get a man to understand something when his salary depends upon his not understanding it." Most salaried journalists are just mockingbirds regurgitating lies that they have been fed by their Lord of the Lies.

Are we doomed? Who will save us from the impending plague and pestilence? Who will lead us out of the desolation into the Promised Land?

"We are the only ones who can save ourselves, but most of us will not take the red pill," said David, "I for one will never take it. We may complain a lot but we're too comfortable with the lies. We are complicit to all the atrocities brought about by this good thing we call money and all the lies that come with it. Besides, why would I want to join the clueless, craven, squabbling, spineless, hypocritical, delusional, and depraved laptop rebels who revel in attacking each other rather than the problem? Nothing will change until we acknowledge that we are the problem, and then someone can do something about it. But not me—I have a family and a mortgage."

CHAPTER 18

Tevye's Conundrum

The scales had fallen from Victoria's eyes. The truth isn't kind to the ear; it's in fact very hard to swallow. The powerful have always taken what they want from the defenceless and they have always made good moral excuses for it. They own everything, including the narrative. Despite their righteous advocacy for protection of property rights, they're really talking about their rights to own property that they have plundered from others. Then they'll righteously vilify those they have robbed to be lazy parasites and thieves. The peasants have to accept it, even embrace it, or they will be snuffed out.

"Thou dost not own anything that thou canst not defend or hide," said David. "This is the first principle of ownership, and therefore the first law of economics."

"I'm sorry I can't agree with some of the things you're saying," exclaimed Victoria. "Should we follow the ancient laws of the jungle?"

"It's the law of nature," David said. "All living organisms and their economies must obey it. We all compete for limited energy resources and use that to reproduce, adapt, and develop. Sometimes we compete; sometimes we cooperate; mostly we do both at the same time in a balancing act, like a fiddler on the roof. When we balance well, we thrive and survive; when we lose balance, we fall and die. Nature is merciless. Lao Tzu (老子; lao-zi) said it, and you can take that to the bank."

"Should we act like animals then?" asked Victoria. She wasn't sure what David was advocating.

"Quite the contrary; if we are the master of the earth, then we'll have to be responsible for it," said David. "With great power comes great responsibility. We'll have to maintain balance in the predator-prey relationship. The ancient Chinese sages had specific laws recorded in the *Book of Rites* forbidding the wanton exercise of power. Confucius (孔夫子: kong-fu-zi) also has a simple adage which says, 'Control your desires and be respectful. That is to be saintly.' There is only one result for people who ignore this law—self-destruction."

"I get the idea," said Victoria. "Don't kill all the farmers, and keep your food safe. I learned that when I watched the Seven Samurai."

"Hear, hear," exclaimed David. "But let me conclude with the second law of economics: 'Every economic ism has its truth but they only work for certain situations some of the times. They're not mutually exclusive and they're not infallible.' This is known as *Tevye's Conundrum*."

Who is Tevye? Is he another one of David's whimsical lie? Tevye is the milkman and the long-suffering father of four daughters in the stage musical *Fiddler on the Roof*. In a scene at the market, one fellow makes a statement; Tevye says he is right. Another fellow disagrees; Tevye says he is right. A bystander protests, "But they can't both be right." Tevye says, "You are also right." That is *Tevye's Conundrum*. Everyone is right, but conditionally, partially, temporarily, and almost never universally or exclusively. Reality is complicated and it's always changing, like a unicorn. No single economic theory encompasses the whole truth or the only truth exclusively and eternally … except for this truth, which is why it's a conundrum.

"Then according to *Tevye's Conundrum*," said Victoria, "there is such a thing as a free market."

"Of course," said David, "but not based on the lies we know."

To elucidate the mechanisms of a free market trade, David told the story of the first free trade in human history.

One hominin has a pile of apples but no stone axe. Another one has two stone axes but no apples. In order to satisfy each other's need, they trade. The fruit hominin gets an axe and still has half a pile of apples. The axe hominin still has an axe and now has some apples to eat. Once the hominins learn that they can trade their surplus to get what they need or desire, they develop specialization, followed by science, technology, art, and civilization. That's what we learn from the books.

In the real world, the hominin with two axes sees an unarmed hominin standing there alone with a bunch of apples. A problem arises. The first issue of trade is ownership. You can't trade things that don't belong to you. The axe hominin wants proof of ownership. The fruit hominin says, "I picked them from the trees." The axe hominin says, "You don't own the trees; therefore you don't own the apples." In fact, no answer is satisfactory to the axe hominin, who says, "God provides us with the apples. You can't prove you own them. I'm just going to take what I want."

The fruit hominin notices the axe hominin has his hand on his axe, so he meekly lets the axe guy take some apples. It's either his apples or his life, a fair deal that he can't refuse. The fruit hominin thinks, "Next time, I will get my brothers to sell fruits with me, and I'll keep an axe at all times." This is what trade looks like in real life. No one will respect your ownership of anything if you cannot defend it, and you can't trade what you don't own.

"Free markets, free speech, free press; they all exist but not without limitations," said David. "Free is just an adjective. Be careful about adjectives being bandied around by self-righteous demagogues who probably don't know what those words mean. But regardless of what adjectives you use, good markets are those that maintain a balance between violence and trust. Smart business people understand that too much violence is bad for business, and it's almost impossible to do business without trust. I'll have to strongly disagree with the capitalist lie that selfishness and greed are necessary virtues for human progress."

"I think selfishness and greed are natural instincts, but they shouldn't rule us," Victoria said. "What if mothers look at their nursing babies and say, stop, you're eating my lunch? We won't survive past one generation."

"That's right, a selfish person has no supporters and therefore will have a hard time defending his ownership of property," said David. "A successful society is built by people who have trust, who help each other, defend each other's rights, share each other's successes, and who can maintain a delicate balance of public benefits against each person's reasonable desire for self-preservation. There needs to be a balance. I'm not saying Adam Smith and Gordon Gekko are completely wrong, but they do not represent the only truth."

"Of course," Victoria concurred, "*Tevye's Conundrum* rules."

David went on to tell Victoria where he got his economics enlightenment, which was an amazing story by itself. In his final year at Peking University, David sat through a course taught by a professor who was the first Chinese national to receive a doctorate in economics from the University of Chicago, the breeding ground of Nobel laureates, especially in the field of economics.

David learned much later that the professor was a ROC military man from Taiwan, who defected to the PRC at the mainland in 1979. At a time when China's average income was about a third that of sub-Saharan Africa, and when Chinese people were trying to get out of China, Professor Justin Lin Yifu (林毅夫)

risked his life by swimming about two kilometres in the dark of the night to mainland China. The Chinese government kept his defection a secret until his wife and kids were able to join him.

While a young man at Taiwan, Professor Lin was already a well-known poster boy of the Republic of China (ROC) military. He was happily married, and had a two-year-old son with another baby on the way. He was also a native son of Taiwan with an extensive clan on the island. His family had been in Taiwan for many generations. He risked losing everything to become a poor man in a poor country with no name and no family. Years later, the farsighted professor would be appointed Chief Economist and Senior VP of the World Bank. China's economic miracle over the past forty years happened as a result of hard work, sacrifice, and in no small part, economic policies laid down with the help of sages like Professor Lin.

"But if everyone lies about everything," asked Victoria, "should we even trust ourselves?"

"Certainly not. We constantly lie to ourselves," said David. "If we admit for a second our tacit or active complicity in atrocities committed with our hands and our wealth, sometimes with our blessing, we will not be able to live with ourselves."

"I don't think my parents and I," exclaimed Victoria, "have ever treated anyone poorly."

"You're still young, but you should remember what I'm telling you today," said David. "People plead ignorance, but it's mostly wilful ignorance. We are all hypocrites, and hypocrisy is a two-headed snake; it bites the hand that feeds it. Besides, our senses lie to us all the times; it's a matter of survival. You're a fan of *Amazing Randi* and *Penn & Teller*; you know how magic takes advantage of this human weakness."

"Never trust a magician," said Victoria, "that's what they preach."

"How true," exclaimed David. "And to wrap up this dialectic on lies, let me quote something from a professor."

David browsed the car's built-in iPad.

"Here it is, from Professor Harari's book *Sapiens*: 'Myths and fictions accustomed people, nearly from the moment of birth, to think in certain ways, to behave in accordance with certain standards, to want certain things, and to observe certain rules. They thereby create artificial instincts that enabled millions of strangers to cooperate effectively. This network of artificial instincts is called culture …'" David quoted. "Well, myths and fictions are but nice names for lies. We have artificial instincts—a polite way to say we're brainwashed since birth. So, there we have it. A learned academic, basically agrees with my claim that everything is a lie."

"What do you call this?" Victoria asked, as she showed David her half-full cup of hot chocolate. She suddenly flipped her cup over on the little table in front of her. David was taken aback. Fortunately, no hot chocolate splashed out. Victoria lifted the cup to reveal David's Breguet tourbillon watch. David looked at his wrist astonished, laughed, and applauded with a "Brava" as he put his watch back on.

"That's awesome magic," said David. "It may come in handy someday if you find yourself surrounded by hostile savages."

Victoria thought of the happier times just a few days ago when she performed magic tricks and the *Cup Song* for her parents. She started to do the *Cup Song* with the paper cup.

"I got my ticket for the long way 'round, two bottle a whisky for the way, and I sure would like some sweet company …"

Harmony and balance returned to the universe as music filled the void of outer space. The shadow of a phantom vehicle sped through the darkness on a deserted highway somewhere in the Canadian prairies. Up in the sky, the Orion shone bright. A cigar-shaped interstellar rock tumbled and rolled across the *Orion's Belt.*

CHAPTER 19

The Watchers Cometh

The news of the accident that wiped out Victoria's family became common knowledge in the neighbourhood by the next day. The police was able to locate the license plate, which flew off in the explosion, but the firefighters could not put out the flames until it had incinerated all traces of human tissue. After the lightning struck, all the electronics in the vicinity experienced complete failure, further hindering the rescue action. By all accounts, the whole family perished in the unfortunate incident, and the police made a public statement to that effect.

Jackie and Bella were worried sick about Victoria being on her own while threatened by unknown killers. When the girls got the message "IMOK, OMW2PVG, ⚙KIT" from Victoria, they were ecstatic.

The next morning, on Christmas Eve, Bella went to Jackie's place after breakfast to decipher the message together. It's fairly easy as the *Weird Sisters* had been doing this for some time. IMOK was a standard phrase not meant to be obscure. OMW was another standard code for On My Way; but PVG needed a little work. Jackie had flown quite a bit and noticed that it looked like an airport code. Googling airport code PVG gave the answer Shanghai Pudong International Airport. They should have used an untraceable search engine such as the DuckDuckGo. This minor slip almost destroyed our universe.

KIT was a standard code meaning Keep in Touch. But what is the meaning of ⚙ ?

“It looks like a gear,” said Jackie.

“It could be a wheel.”

“It is a wheel,” Jackie laughed. “The whole message is: I’m OK; on my way to Shanghai. Wheel (will) keep in touch.”

They gave each other a high five. Jackie promptly sent back a happy face emoticon and deleted the messages.

“My mom told me that a couple of detectives came by the house last night to ask some questions,” said Jackie. “She didn’t think it was anything out of the ordinary. But we should alert each other if we see anything suspicious.”

When Bella got home, she saw a black BMW SUV parked on the driveway. She decided to get in from the back unannounced and unnoticed. As Bella sneaked past the living room, she saw two men talking to her parents. They were wearing long black overcoats, black pants, black leather shoes, and black gloves. One of them carried a black leather briefcase in his hand. They had their back towards Bella so she couldn’t see their faces, but she had a bad feeling about this. After they left, Bella came down from her room to talk to her parents about the intruders.

“What was that about?” asked Bella.

“It’s great news,” said Bella’s mom. “We have just won a prize for the whole family to visit Disneyland, all expenses paid.”

“Really?” cried Bella. She couldn’t believe her ears. “You’re kidding me.”

“We have to go right away,” said Bella’s Dad. “Pack some clothes quickly. We have to be out of here in half an hour.”

This seemed super weird to Bella.

“Let me call Jackie to tell her about it.”

“No need to do that. Their whole family is coming along,” said Bella’s Dad. “This is no joke. We’re all going to Disneyland.”

Chapter 20

#1 Corn Gas

Two men, both draped in a long black coat over their black suit, black silk tie, black cotton trousers, shiny black leather Oxford shoes, and black deerskin cashmere lined gloves, stepped out of their black BMW SUV and took off their futuristic black sunglasses. They cast a stony stare at the lonely #1 Corn… Gas service station in the middle of the Canadian prairies. One of them held a mysterious briefcase in his hand. The other man, an African American, stooped to pick up the broken GPS/Altimeter watch. He held the gadget in front of his squinted eyes and examined it in ominous silence. He turned to his partner who remarked with a scornful sneer. Then they stepped inside the gas station.

Brent the attendant was absorbed in his comics as usual. The doorbell broke his concentration. He sighed and put down the book, looked at clock, and tried to sound friendly to the undesirable business just before closing.

"Gentlemen, we're closing early today," said Brent. "It's Christmas Eve."

"Do you know who we are?" asked Vincent.

"Hmm, let me guess," Brent responded with a touch of his customary sarcasm, "you're the *Men in Black*, here to deal with aliens on earth."

Brent wasn't amused with people asking the wrong questions in the wrong place at the wrong time. "Listen, fellas," he said, "it is closing time. We're closed, capiche? My beautiful sexy wife is waiting at home with a bird in the oven and a glass of mulled wine to warm the cockles of my heart. So why don't you gentlemen make like a tree and take your business to the next service station about a hundred klicks (sixty miles) from here? You're welcome."

Brent kept his uneventful days interesting by being sarcastic, but he chose the wrong targets today.

"You may want to know who we are," grunted Jules the humourless African American through gritted teeth.

"Not really," said Brent. "I think this whole episode has something to do with one of my friend's crazy daydreams or I'm in the middle of a Hollywood movie."

Brent glanced at the security camera as a signal to the strangers not to try anything funny.

"Answer our question and you can go home to your sexy wifey," said Vincent as he imposed himself in front of Brent. "Did a man of late forties early fifties and a teenage girl, both Chinese, pass through here recently?"

"Why am I talking to you?" Brent said, remembering the hundred-dollar bill he received for keeping mum. "You're not getting gas or buying anything. What's in it for me?"

"Fine, I'll have the Royale with Cheese and my brother here wants the Big Kahuna," said Vincent, matching Brent's sarcasm with one of his own. He then pulled a large pistol from his coat and almost stuck it up the nose of the shocked attendant.

"Look, Rockefeller, this is what's in it for you," said Vincent. "If I'm not satisfied with your answers, I'm going to rip your face off with my bare hands, then my friend here (wiggling his gun against Brent's nose, making his nose wiggle) will put so many holes through your cold dead body they'll have to wipe your smugly face off the floor with a mop."

“Well, why didn’t you say so right away? You should’ve said so right away,” said Brent as he made a violent sneeze. Jules said, “Bless you,” and Vincent said “Gesundheit.” Brent blew his nose, said, “Thank you,” and complied with the wishes of his guests. “I’ll tell you everything you want to know—my first kiss, my secret chili cheese dog recipe, even where I hide my girlie mags.”

“Do you read the Bible, Brett?” asked Jules with a dark menacing growl. He didn’t like where the interrogation was going.

“The name is Brent,” said the bewuthered gas station attendant, pointing to his nametag.

“No, not the Bible again!” Vincent protested as Jules prepared to recite some apocryphal Biblical verse before blowing someone’s head off, “You made that whole thing up.”

“I most certainly did not,” asserted Jules. “Ezekiel 25:17. Look it up.”

“You want to make a bet?” said Vincent with a smirk on his face. “If I’m right about you making that up, are you going to give me that foot massage?”

Jules turned to give Vincent the Doonesbury stare with his cold dead eyes for about five seconds and then blurted out, “Figaro! Are you trying to piss me off?”

Jules then pulled out a huge Smith & Wesson 500 Magnum from inside his overcoat and pointed it at Brent’s face. “Talk!” he growled.

“I d-don’t know anything,” said Brent, obviously intimidated by the gnarly blunderbuss. “W-What do you want to know?”

“Tell us everything,” Jules said.

“W-What do you mean everything?” said the poor flustered Brent.

“EH-VRY-THING!” yelled Jules at the top of his lungs.

“W-When I was four years old,” said Brent, “I remember the time my m-mother tried to toilet train me …”

"About the Chinese man and the teenage girl," thundered Jules, while glaring at Brent with his enlarged eyes and contracted pupils.

"Last night, a middle-aged man and a teenage boy," said Brent, "both Asians, could be Chinese, Koreans, Filipinos—I don't know—came in around closing, got a drink and left. That's all I know."

Brent finally sang like a canary. He didn't think it was wise to fool around with these two goons anymore.

"By the way," added Brent, "there was no teenage girl."

Vincent and Jules looked at each other.

"Are we happy?" Jules asked Vincent.

"We're happy," replied Vincent.

Chapter 21

The Unicorn Rings

Lady Diane had summoned the commanders of the various units to Queen Zia's tent for the war meeting. The queen's army was the most sophisticated military organization of the times, and Lady Diane developed it. Shang hoplite order, which was the best in town, was organized in units of ten, a hundred, five hundred, and a thousand. Depending on the number of chariots available, twenty-five to fifty men fought as a unit under a phalanx commander in a chariot drawn by four horses. Lady Diane took the structure and trained specialized fighters for specific tasks. But for the coming battle, Lady Diane sensed that the Gotts were not going to be easy prey. They had their backs to the river, so for them it was a case of do or die. There was no retreat—a cornered beast is a dangerous beast.

Contrary to the common belief that the Gotts were wild savages, they actually had an organized army with a division of armoured cavalry and another of mounted infantry. They used the grain wagons to form a mobile fortress where they tied up all of their captives, most of them young men and girls from the Shang villages. Would Queen Zia burn her own grains and cause the death of her own subjects?

On the other hand, Lady Diane had planned to spring a few surprises on the Gotts. Fighting would start early in the morning with the sun shining into the Gotts' eyes. Early battle would also surprise the Gotts. The Shang armies always made an elaborate offering to the sun at first light. Then they ate a full breakfast. After the battle lines were drawn up, the military augur would ask the oracles whether to fight or not to fight. It's surprising that the Shang armies ever fought any battles at all.

Lady Diane wasn't going to follow unreasonable rules. She sent camp heralds around the camp announcing the favourable oracle for early battle and the promise of rewards. Aside from the normal victory bonus, for each enemy's left ear collected, the reward would be ten cowry shells. For killing an enemy chieftain in single combat, the Shang commander would get the spolia opima (rich spoils), which included all the arms and properties of the slain enemy chieftain. In addition, the queen would reward the commander with ten bronze cowry shells, and one bronze spade known as *Qian* (钱; money). The commander would be a hero and he would be rich. Disobedience and cowardice during battle would be rewarded with instant death.

Just to clarify that the soldiers weren't getting trumpery as their battle bonus, the cowry shell was a currency of the Shang Kingdom. The cowry shells were beautiful shells from the oceans far to the southeast. The word for 'cowry-shell' (贝: bei; OB: ) is an important radical (root) which forms a part of many modern Chinese words that relate to money and trade. The Shang people were famous traders—some say they were the first merchant-traders in China. Their Promethean ancestors brought the cowry shells along their migratory route from the south Pacific coast. When the cowry shells became worn, debased, and inadequate as society developed, bronze money appeared.

In the early days of Bronze Age in China, any useful and valuable bronze object that required a lot of expertise and efforts to make became money. As we have learned earlier, without peasants to produce the surplus food, there would be no money, nothing to trade, and only primitive civilization.

The bronze spade *Qian* was valuable and became currency. At first, real bronze spades represented a huge sum of money. Later people minted tiny bronze spades for daily trading between individuals. Eventually, the bronze spade called *Qian* became the Chinese word to mean 'money.' When a Chinese asks if you have any money, he is actually asking if you have any spades. And if a Chinese lends money, he's going to call a spade a spade, and he'll want to be paid back in spades.

After the meeting was over, Queen Zia told Lady Diane to stay with her for the night. She then laid out a pair of beautiful jade archer-rings in front of Lady Diane. A one-eyed trader from beyond the Celestial Mountain in the northwest had given these rings to the queen. He wished to gain the audience of the king to apply for a license to trade in the capital. Bribing the powerful queen was a clever way to go about it. The trader would eventually settle in the capital Yin (殷; thriving) and later became an important official of the Shang court. He called himself the one-eyed official Huang Yin (黄尹).

Queen Zia granted Lady Diane the first pick. Lady Diane chose the one with the words ( ), which meant 'return-not,' leaving for Queen Zia the ring with the words ( ), which meant 'go-come.'

"Use it for the battle tomorrow," said Queen Zia. "And may every arrow you shoot find its target."

Cowry money

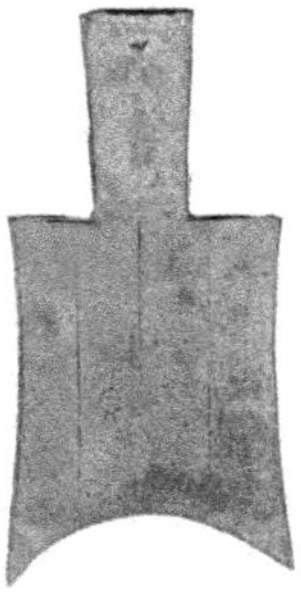

Spade Money

CHAPTER 22

The Oracle

Clever AI assistant Moira scheduled a restroom stop and a midnight snack for the two Chinese Canadian fugitives from Toronto.

"Are you who you say you are?" asked Victoria as she munched on her favourite Canadian snack at the service station. Victoria loved French-fries with cheese curds and gravy—the unique Canadian culinary invention known as poutine. Belgians of course beg to differ, because aside from declaring that they make the best beers and the best steaks, they also insist that they invented the original fried potatoes which they call Belgian fries, and which they serve with all kinds of awesome sauces.

Victoria craved poutine and called her incurable habit the Poutine Routine. Lucky for Victoria, China had everything including poutine. Canada's leader known in history as Trudeau the Younger went to Shanghai in one of his visits and had a poutine fix at Tock's, a popular Canadian restaurant in the city.

"I'm very much yours truly David Huang," said Victoria's guide and protector. "The symbol () you see on your father's letter and on my ring is the ancient word for my family name Huang (黄), meaning the colour yellow. It's a word borrowed from the name of an arc-shaped jade ornament. Royalty in ancient China wore it as a symbol of their status. You have such a *huang* (璜; modern word adds a jade radical) jade pendant."

Victoria put her hand on the pendant to make sure it's still there.

"The word *huang* meaning yellow is an important word in China," said David. "The dominant Chinese culture was born along the Yellow River—Huang He (黄河). The first mythical emperor of China was called the Yellow Emperor—Huang Di (黄帝). The part of the ocean to the east of China and south of the Korean peninsula is called the Yellow Sea—Huang Hai (黄海). The list goes on and on."

"Chinese doesn't seem that hard to learn," said Victoria. "The yellow school bus is a *huang* bus."

"Correctamundo!" said David. "Before understanding more about yourself, you should learn a little bit about my background. I came from a place in China's Shandong (山东; mountain east) province called Yellow County—Huang Xian (黄县). It is now a part of metropolitan Long Kou (龙口), which means Dragon's Mouth.

"It was at my ancestral home where I first saw you as a newborn baby. My father, whose name is Huang Shi (黄石) meaning yellow-stone, brought you to our house. My father told me that you were a princess, and ordered me to protect you with my life. I know this sounds crazy, but there is a very good reason why I believe him."

"How come I've been kept in the dark?" Victoria asked. "Are my biological parents from Yellow County? Are they descendants of some ancient dynasty?"

"The secrecy is needed to hide you from bad people," said David. "The only person who knows any details about your real parents is my father, but he fell into a coma before he could tell me everything. You may be the only one who holds the key to the secrets in his head. That is why I'm taking you to him."

"I'm sorry about your father," said Victoria, "but this is a wild goose chase."

"Don't despair. You will understand why in a moment," said David. "But we'll start with the crazy story of how and why my family and others got involved in protecting you."

Believing that he had somewhat softened Victoria's resistance to the ludicrously implausible, David began to consolidate his gains with some well-known historical facts that were related to Victoria's background.

While China has about four thousand years of proven history starting with the Shang dynasty, modern Chinese people have mostly lost the genealogical connections with their ancestors. Other than some nebulous unprovable claims of ancient lineage, most Chinese nowadays don't know their family history beyond a few hundred years. There are a few notable exceptions, one of them being the lineage of Confucius.

The Kong (孔; originally meant good, family name of Confucius) clan has an unbroken patrilineal record extending for twenty-five hundred years going all the way back to Confucius himself. Confucius has many famous descendants throughout history. One of the most famous Kongs in recent years was the 75th generation descendant Kung Hsiang Hsi (孔祥熙: kong-xiang-xi), better known as H. H. Kung, China's famous financier before the Communists' ascension to power. He was the brother-in-law of Dr. Sun Yat-sen and Generalissimo Chiang Kai-shek (蒋介石; jiang-jie-shi). Dr. Sun was the founder of the Republic of China (ROC). Chiang took power after Dr. Sun's death, ruling China until he lost the civil war to the Chinese Communist Party (CCP) in 1949.

H. H. Kung married the eldest of the three Soong sisters Ai-ling. Dr. Sun married the middle sister Ching-ling, and Chiang Kai-shek married the youngest sister Mei-ling. The three sisters at one time were the most influential women in China. Some people remarked that Soong Ai-ling (宋霭龄: song-ai-ling) loved money, Soong Mei-ling (宋美龄: song-mei-ling) loved power, and Soong Ching-ling (宋庆龄: song-qing-ling) loved China.

Victoria knew about the three Soong Sisters, but she didn't know that they were related to the most influential family in China, namely, Confucius's family. Was that coincidence or was that six degrees of separation? Victoria didn't know any Chinese history. She thought, maybe the next thing David was going to say was that she was descended from Confucius.

David Huang told Victoria that his Huang family had a secret unbroken lineage that extended for thirty-two hundred years, about seven hundred years longer than the Confucian clan.

"Dear me," thought Victoria, "and I don't even know my own parents."

According to David, this lineage record, kept for a hundred and eight generations, stated that the clan's first ancestor and founder was a high level court official during the Shang dynasty by the name of Huang Yin (黄尹). Yin (尹) is an official title, sort of like Minister. The ancient Shang Oracle Bone word for *yin* (尹) is the drawing of a hand holding a walking staff (). Huang Yin's name means 'Aristocratic Minister.' David's lineage record stated that he was a one-eyed official of the king.

David discovered from Shang archaeological records numerous Oracle Bone inscriptions that attested to the life and deeds of Huang Yin. Since the oracle bones had only surfaced in the last hundred years, David proved without a shadow of doubt that his secret lineage record was authentic, and that his clan's mysterious ancestor was a historical person, not a fabrication.

"Before I found the Oracle Bone records, I was just as skeptical as you are," David told Victoria. "There are in fact many secret clans of long unbroken lineage, not just my own, and not just from China. For example, Raja came from such an unbroken lineage within the Bhargava clan in the northern Indian state of Haryana. All of us have been keeping a secret for generations. I'll tell you what the secret is but you'll have to first swear to keep it a secret."

Victoria assured David that his secret was safe with her, and that even if she told anyone else, they would just think that she was totally bonkers. David therefore revealed the first piece of Victoria's puzzle, which linked her to people and events of more than three thousand years ago in China.

The Confucian clan might have been around for twenty-five hundred years, but Confucius was also the scion of a secret clan that had begun much earlier. Confucius was a bastard son, as his parents did not officially marry. His father already had an official wife, a concubine, and a firstborn son. Yet Confucius inherited the secret lineage book from his father, who died when Confucius was three years old. His father did not want to pass the all-important secret to his firstborn, who was physically and intellectually challenged. It also proved that the secret was more important than the bloodline or legitimacy based on primogeniture. The legitimate Kong clan kicked Confucius and his mother out of the Kong household immediately after his father's death. Confucius went on to establish his own lineage and kept the secret clan going as well.

Confucius was no common bastard. He had the royal blood of the Shang in his veins. Legend said that the Chinese unicorn known as the Qilin (麒麟) appeared at the birth and death of Confucius. In the unicorn's mouth was a jade piece, and on the jade was the oracle about the arrival of a king, one without a throne or a kingdom.

The secret lineage book has a cryptic password by which the clansman privy to the secret may contact other secret clansmen upon the revelation of a certain sign. The book also states that the keeper of the clan secret must protect the one who bears the oracle of the stopping-of-the-sun. When the oracle appears, it will be the end of the world. The bearer of the oracle is the one who will save humanity from extinction. When Confucius was born, and word went around that he bore the oracle, the secret clans thought that he was the one, but it turned out he was important in other ways.

"For over a hundred generations, the secret clans waited but all we got was silence," said David. "Many clansmen have given up, until you come along. You have jade pieces with the oracle of the stopping-of-the-sun. We have no doubt that you are the one."

"What is the stopping-of-the-sun?" asked Victoria. She had a hard time believing any of this was serious. "You mean, like when the earth stops spinning and everyone dies? Or is it a warning from advanced beings like in *The Day the Earth Stood Still*? Am I the one who'll reveal the seventh seal?"

"The stopping-of-the-sun sounds ominous, but it's not what you think," said David. "Let us get back on the road and I'll explain on the way."

Revelation: The Sixth Seal by Dürer

Chapter 23

The One-eyed Monster

WARNING: The following is a factual story that is extremely disturbing and may cause psychological trauma. Although events such as these occur every day and everywhere, some of which we may even be party to, we are normally shielded from them by our benevolent press and by our own self-defence faculties to ignore and forget. For those who need protection from the terrible truth, please avert your eyes and ears, skip to the next chapter, lay down the book, fall upon your knees, and pray to the gods to deliver us from evil. Parents should make sure children do not read this chapter without supervision. You have been warned!

Any casual observer knows that the universe is mostly peaceful. The violence of creation and destruction is local and temporary. Nature is like a pendulum, which when disturbed swings from one extreme to another, but ultimately settles down to a resting equilibrium until it is disturbed again. Peace will return. Peace always returns. Santos Didymas said so.

In this vast peaceful galaxy, the vision of a handsome young boy came into view. He glowed like a phosphorescent apparition in the dark of deep space. He loved the galaxy, he loved the stars, but most of all, he loved Shangala, the planet he called home.

In this serene and peaceful dream, the star child heard only music. This was not ordinary music. This was the harmony of existence, the beauty of symmetries, the art of fugal balance, and the miracle of self-emergent order. While peace reigned, existence was wonderful, beautiful, and inevitable, and it was expressed in the ever-changing colours and infinite musical tones of Shangala.

Without warning, the star child burst into flames. Even in a dream, he could hear a terrifying scream—it was his own. People say that in space, no one can hear you scream. That is not true. You can hear yourself scream because sound travels very well along your bones. And as sound and pain travelled along his nerves and bones at the speed of light, Kiku woke into cruel reality.

Kiku's parents belonged to the caste of musicians in the Haryan hierarchy. They did not produce food, they did not produce tools, nor did they invent technologies for war. They were useless fools. Even within the caste, musicians were separated into several classes. The top rung made music for the rulers and the bankers; the next rung made music for the henchmen and the arms-makers; the third rung made music for the police and the military; the fourth rung made music to lure and corrupt the enemies; the fifth made music to mesmerize and control the citizenry. Kiku's parents belonged to the lowest of the lowest. They made music because they loved it. They were worthless parasites.

Shangala had been at war for a long time. No one remembered how the endless global war started or why everyone was fighting each other and everyone else. War touched all Haryan except Kiku's parents. They belonged to such a despicably depraved class in the Haryan society that they were not allowed to get involved even if they volunteered. In order to survive, Kiku's parents built themselves a small hut at the edge of civilization using mostly discarded boxes, and there they lived on bare subsistence, while they raised their son and daughter to love music as they did. They were happy to live in relative peace while everlasting war was raging all around them.

When Kiku reached the age of seven and when his sister was four, his parents finally completed the construction of a unique musical instrument. It was like a large round table two metres (six feet) in diameter. They made this contraption completely from discarded garbage and scrap. While the material used was ordinary and base, the design was ingenious, and it followed the design of an earlier instrument known as the Galactic Harp. The shape was round to represent the galaxy. The centre of the harp was Shangala's sun Yin-Di-Ra the Lifegiver, represented by the actual shape of the immortal flower Diamaranth. Harp strings connected Shangala to myriad star systems at various distances from the centre. It was a harp based on Shangala's star chart.

A Yarjun apostate known to posterity as Rock Strongarm had invented the principles of the Galactic Harp many centuries ago. He built the original instrument with the wood of the Deathrider, the mythical giant tree that was so huge its roots encompassed the entire planet. Rock Strongarm wanted to bring peace to the world with this harp, but the Yarjun atomized him and his musical instrument with a homing plasma missile from an armed drone. After that, no one dared to repeat the foolish act until Kiku's parents. They figured they were so insignificant and so distant from society that no one would ever pay attention to their private worthless music.

When Kiku's parents played the Galactic Harp for the first time, the resonance of the Deathrider's roots deep inside Shangala caused the entire planet to hum. People all over the globe heard these strange sounds. Everyone on the planet stopped what they were doing to listen. Even the Endless War took a break for five minutes. Unfortunately, people were unaccustomed to peace. After realizing that it was only music, they promptly resumed the mutual slaughter.

It was a night just like every other. Kiku and his sister slept in a large round bed shared with their parents—the kids at one end, the parents at the other. The room was small but it felt like they had the entire universe to themselves. They had a nightlight that created a holographic galaxy in the room. Dream music composed and performed by Kiku's parents played from a memory device. It lured them into a dream world where Shangala was still innocent, and murderous simians had not yet declared ownership of the planet.

When Kiku woke up, he could smell the intoxicating aroma of burnt flesh. His body was in flames—a beautiful blue blaze known as the Fires of Ork, a cruel weapon accidentally invented by Yosi the Toymaker. The blaze was non-lethal, but it would torture Kiku and disfigure him forever. Other than the sound of his pain, Kiku could hear nothing else, not even his most favourite *Intermezzo*, which was still playing from the memory device. Kiku would never enjoy music the same way again. His ear drums had been ruptured by the explosion.

The hut's ceiling was gone, exposing the peaceful night sky and the three pale red moons of Shangala, all at different phases. The only sign of malicious mischief was the network of contrails left behind by the squadron of drones that had lurked in the neighbourhood. Through the dust and the smoke of the explosion, the moonlight showed Kiku an evil he wished he did not have the power of sight to see. Someone had dropped a guided cluster bomb into his home. It was banned ordnance but the law was cold comfort for Kiku, because the perpetrators would be hailed as heroes tomorrow on the Freedom and Truth Media.

The bomblets were fiery spinning scythes meant to lop off any body parts that happened to stand in the way of their pitiless paths. His parents lay motionless in bed in an endless dream of death from which they would never wake. Their heads were cleanly severed from their shoulders. His sister beside him was spared the agony of life, for while in her slumber she was bisected across her waist. The sheets had turned black from the river of blood that carried the spirits of the dead into soulless space. The pale-eyed moons morphed into cold-hearted orbs that ruled the night. The cruel countenance of death would forever haunt the young boy's sight.

The universe did not create enough tears for this innocent lad of seven, and all of it gushed forth from his one remaining eye. A steel ball had crushed his left eye into a crater of blood. Kiku wanted to wipe his blood away but he couldn't. Looking through the teary window of his tortured soul, he realized that the spinning scythes had sliced off all his limbs. He had only stumps where he once had agile arms and legs. The flames that had engulfed him inadvertently cauterized his wounds and saved him from bleeding to death. He wished the merciless scythes had taken his head as well. Why was he still alive? Why did he have to suffer life? What was the point of it all?

In his despair, Kiku swore by the lifeless bodies of his parents and his sister, that if he survived, he would avenge their deaths without regard for cost or pain. It would be a lesson to the Haryan, and a most important lesson for their survival as a species, that criminals who committed atrocities would not enjoy impunity.

No one understood why anyone even bothered to bomb the musician's home. His whole family was less than insignificant. It was a waste of money and munitions. Perhaps the bomber used the wrong map. But no one really believed such a bald-faced lie except for the gullible citizens of Hypnerambomachia and Poliantia. No one claimed responsibility. No one wanted to investigate. It would be throwing good money after bad.

The hospital kept Kiku alive because he was now a valuable asset. In the ensuing bidding war, the Yarjun offered the highest price and made the most serious threat to slaughter everyone at the hospital, including their extended families and their electronic pets, if the Yarjun's offer was not accepted. It was an offer the hospital could not refuse. They comforted themselves that it was a fair deal. Under these circumstances, Kiku became a ward of the Yarjun and the focus of their propaganda war.

The Yarjun gave Kiku bionic limbs. They replaced his left eye with a *Visual Spectrum Speculator* (VSS). They grafted polymer skin all over his body and face. They built him a hut that resembled his old home, and they supplied him with androids that acted like his parents.

The Yarjun did everything they could to instill true gratitude in the impressionable young boy. They put Kiku on the news every day. They condemned their enemies incessantly, and repeatedly vowed to investigate. Their pundits praised the courage of Kiku. Reality TV camera crews followed Kiku everywhere all the livelong day. Their media and entertainment industry produced television specials, docu-dramas, movies, and stage musicals about him. Fable-mongers known as logopoists wrote epics and sagas about Kiku's ordeal. Poets and bards composed poems, psalms, and songs to extol his heroics. Kiku had become a modern day legend before he turned eight.

Donations, endowments, and prize money poured in. Kiku's Yarjun benefactors gave him the best education possible. A war erupted between factions fighting for Kiku to make morality speeches at their Indoctrination Institutes. He signed autographs at annual Black Death Sales events, sang Yarjun Torture songs at Blood Sports arenas, and crushed skulls at warcraft launches.

On a glorious day commemorated by a national holiday and a parade, the state of Yarjun officially adopted Kiku. They elevated Kiku from the lowest musician caste to the highest Yarjun warrior caste. They even gave him a Yarjun name—Deskull, a name befitting a Yarjun Class-One warrior, national champion of Slasher War Games, world champion of multidimensional multilateral Blood Chess, world record holder of reality Slaughter Games, and proud owner of eleven patents for innovative war concepts, the greatest being the *War on War on War*. Deskull would become a master of his craft. He would go on to train many of the Supreme Lord Killiam's frontline fighters.

The Yarjun at one point became the paragon of virtue in the minds of all who swallowed the lies. Their popularity polls went through the roof to the 80's percentile. Their investment in Kiku was repaid in spades. It seemed inevitable that Lord Killiam was going to become the Lord of Shangala, until one day, Deskull did not show up for work. Deskull had deserted to join the universally despised Legion de Lude, to march with the Animorta soldiers. The Yarjun then began an unstoppable fall from grace.

Deskull had accidentally come across an archival record of the Yarjun and learned who had dropped the guided cluster bomb on his family. The Yarjun feared that the Galactic Harp would cause the Haryan to stop fighting. Peace would destroy the Yarjun caste. The existential threat of peace must be eliminated with extreme prejudice.

It was supposed to be an easy covert decapitation operation—literally—on a defenceless and unsuspecting civilian home. No one would give a rat's ass. The victims would be bugsplat. The Yarjun had not expected anyone to survive. Converting Kiku into one of their own and using him for propaganda was a stroke of genius, turning failure into success and defeat into triumph.

Although the Kiku file was a classified secret protected by the highest level of confidentiality, the Yarjun leadership forgot to hide it or destroy it when Kiku became Deskull, who would rise through the ranks, and eventually received the highest security clearance. Deskull went to the archival library to study historical cases of Yarjun war deception. By accident, a thin file fell on his lap. Seeking lies, Kiku discovered the truth instead. History was changed because of one little mishap.

Chief Deskull had defeated the Yarjun and had transformed himself into the benevolent King Kull. The complete memory of the Yarjun was expunged. It was as if the Yarjun had never existed.

Fear and hate had blinded the Haryan of Shangala, and in the land of the blind, the one-eyed man became king. King Kull had ended the Endless War. He was now the sole ruler of Shangala. His power was absolute and he used his power for good.

King Kull banned war; he banned killing; he banned violence; he even banned hate. King Kull hated the word 'hate,' so he banned it. From now on, all the resources of the planet would go towards rebuilding the paradise lost. By King Kull's order, let there be everlasting peace!

CHAPTER 24

Apocrypha of Didymas

Lord Killiam had underestimated Chief Deskull, whose secret weapon was a tech geek by the name of Millipath, later universally known as Lord Millistar. Millipath was a prodigal genius. When Chief Deskull first met him, they were in every way completely different from each other. They were only united in one dream; that was to end the Endless War and to rebuild the paradise planet of Shangala. It was a beautiful friendship and a perfect partnership. While each needed the other to reach their common goal, neither one could take the place of the other nor would they ever want to.

It was however impossible to defeat the Yarjun because they did not make logical economic decisions in war. The Yarjun would not accept peace under any context, even if it was more beneficial to do so, and they would never accept any reference to the word defeat. The laconic Yarjun had only one proverb in their dictionary: "Defeat stinks!"

They had implemented the Stranghate Option as a last resort. The legendary Dauk Stranghate, one of the founders of the Yarjun Order, designed this strategy of total destruction to terrorize the meek. Unfortunately for Dauk, an unintended consequence of genetic manipulation to increase his martial intelligence quotient a hundred fold caused him to become stark raving mad. They strapped him in a strait jacket and locked him inside a tiny padded cell in Xylem, the hellish dungeon for the criminally insane. They kept him in there for fifty years until he finally figured out how to commit suicide. Dauk Stranghate suffocated himself with violent laughter.

The Stranghate Option stated that, "If defeated, the Yarjun would destroy the planet and let everyone perish with it." Once the Stranghate Option was implemented, the Yarjun would never taste defeat; they wouldn't even smell it. The last Yarjun to die in defeat would trigger a chain reaction that would be unstoppable. It would ignite the Erinysite buried in the core of the planet. Under high pressure, the inferno would reach fifteen million degrees and blow the planet into smithereens.

In order to conquer the Yarjun, Chief Deskull must change their minds. That was Chief Deskull's strategy, but he needed technological breakthroughs from Millipath to make it work. Millipath finally gave Chief Deskull the breakthrough he needed. They have another secret weapon, the unicorns.

Millipath had been studying rare minerals with strange properties, in particular Neodymium and Sandymium. Both were rare on Shangala, Sandymium was rare in the entire galaxy. Both Neodymium and Sandymium belonged to a class of mineral known as Didymium. The founder of the Legion de Lude, sage Didymas discovered this class of minerals, and they named it after him.

In the course of researching the properties of Didymium, Millipath came upon an ancient text known as the *Apocrypha of Didymas* in the library of the Legion de Lude. The Apocrypha was written in an ancient mystical script known as Shanglyphs. No one could understand the Apocrypha until Millipath deciphered Shanglyphs to learn the book's ancient wisdoms.

The Apocrypha spoke of a time when unicorns roamed Shangala. It spoke of unicorns as real animals that had fantastic powers. Unfortunately, unicorns could not survive the rise of the simian creatures. They despised the Haryan, and one by one, they vanished. Even if any unicorns survived, they would hide so well that no Haryan would ever find them.

By the time of Millipath, no living Haryan had reported the confirmed sighting of a unicorn. Most believed that unicorns were mythical. The few who believed unicorns had once lived also believed it had become extinct long ago. Millipath believed in its existence because of the famous saying of Santos Didymas, "Everything is a lie, except unicorns." Santos Didymas also said something about truth, but that's another story.

Millipath followed the clues in the Apocrypha to locate the secret abode of the sage. It was a deep and unexplored cave system of the Dragon Mountain known as the Aurica, which meant 'golden-chamber' in the local language. In the Apocrypha, Didymas called his cavern abode Lasallas, which meant 'home of the nine dragons.' It was the name of Didymas's hometown where the Endless War started. No one knew how it all began. Didymas left his hometown and never returned

It was in the Golden Chamber where the sage first discovered Didymium. It was also deep in the bowels of the Dragon, where Millipath found the last two unicorns, teetering on the edge of expiry.

The Apocrypha said that the unicorn had the ability to meld its mind with other life forms. But the unicorn was also an innocent creature. It would become extremely agitated to detect deception. This was why the unicorns avoided the Haryan, and especially the Yarjun, who made hypocrisy a virtue and deception an art. The unicorns would rather die in a dark dank cave than to breathe the foul air of the Haryan.

By studying the strange properties of the Didymium minerals, Millipath saved the unicorns from extinction. By studying the unicorns, Millipath discovered the most powerful technologies that would give Chief Deskull complete victory over the Yarjun.

From the hide of the unicorns, Millipath developed the new skin for Chief Deskull. His transmutable nano-twinned piezo-amorphous crystalline Boron Nitride epidermis on self-healing 1.1° magic angle twisted multi-layer graphene polymer substrate doped with Tonstone, Yindium, Herolium, Rhinium, and Neodymium made him practically invulnerable to physical attack. This material had the additional advantage of countering extreme exterior temperatures. The unicorn was unharmed by fire because their skin was made of this material.

From the unicorn's eyes, Millipath developed the *Universal Spectral Scanner* (USS) to replace the *Visual Spectrum Speculator* (VSS) on King Kull's left eye socket. Rather than sensing only the visual spectrum of electromagnetic waves, the USS could adapt to a wide range of frequencies with a process known as spectral tuning. It could lock on to a narrow range of frequencies to analyze detailed imaging that would be undetectable by sight alone.

Millipath learned that the unicorns hid deep in the Dragon Mountain for a good reason. They needed to consume Didymium to survive. They needed also to consume Erinysite for energy. Neodymium was rare and therefore precious. The Haryan had been mining it to make powerful laser cannons, electromagnetic missiles, and antigravity engines. Sandymium was extremely rare and had never been discovered anywhere outside of the Dragon Mountain.

According to the *Apocrypha of Didymas*, unicorns would only procreate when conditions for infant survival was optimum. Unicorns would adopt male or female form at the time of procreation. Otherwise, they were asexual.

All attempts by Millipath to induce unicorn procreation failed. The only alternative was to extract the unicorns' genes for cloning. Millipath had developed genetic engineering and artificial surrogacy technology for the Haryan so that people had complete freedom over their decisions for procreation. This technology granted female Haryan full equality with the males. The same technology now gave Millipath a chance to stave off the extinction of the unicorns.

Millipath tried many times with many samples but the embryos would invariably vanish into thin air half way through gestation. Apparently, this violated the Law of Conservation of Mass. After observing gravitational anomalies, Millipath concluded that it was possible for mass and energy to make a quantum dimensional transposition without violating any physical laws. The total energy of the universe was conserved in time subject to quantum fluctuations. It was also preserved in Carlin's Uncertainty Principle as nethermass—phantom material in unobserved quantum states that was neither here nor there—exhibiting dark gravitation only when measured from a relativistic frame of reference.

Carlin was the Haryan scientist who unified quantum mechanics with relativity and scientifically proved the existence of unicorns. Millipath was a genius but he stood on the shoulders of giants.

The Apocrypha said that the horn of the unicorn had a high concentration of Sandymium. The understanding of the connection between the unicorns, Didymium, mind meld, and hyper-dimensional transposition was the critical discovery that would lead to the downfall of the Yarjun.

Despite his power and his prowess, King Kull could not forget the past and the pain. The recurring nightmares of the beautiful blue blazes of Ork left him shaking in his bed, soaked in his own perspiration. King Kull had to calm himself with the Soul Soother, another one of Lord Millistar's innumerable inventions. It was an Artificial Intelligence device that told the most beautiful lies. Only through lies could King Kull heal his tortured soul.

When morning came, the two Seraphanas in the courtyard sang the Flower Duet to usher in the new day. King Kull woke hoping to see his wife Jenni beside him. The sight of her sleeping peacefully and breathing softly always gave him great solace. Then he remembered that she had become a permanent quantized phantom in the indeterminate quantum state.

King Kull suffered the endless pain of a dagger in his heart because he loved Jenni, and she died because of him. He wanted to see his sons and daughters to gain comfort from them, but they were nowhere to be found. He then remembered that for the sake of Shangala and perpetual peace, he had to give them up forever. His old friend Lord Millistar was too busy, working around the clock to finish their final project before retirement—the solution to provide a perfect government and eternal peace for Shangala, the technological wonder that would keep Shangala a paradise forever.

CHAPTER 25

The Odyssey

This was Victoria's odyssey in search for the truth and David was her guide. The Argonauts risked life and limb across the Aegean and the Black Sea in quest of the Golden Fleece. Aeneas followed his destiny through hell and high water, even betraying the tragic love of Queen Dido, to build the Eternal City of Rome. Jeff J. Brown of Oklahoma put on a backpack to trek across China for 44 days to learn the elusive truth. There is no short cut. The reader must tag along for the ride.

David continued with his epic tale once the F011 was back on the road.

"In order to understand the oracle of the stopping-of-the-sun," David told Victoria, "we'll have to decipher the strange glyphs on your *huang* jade pendants."

These picture glyphs are known as Oracle Bone script (甲骨文: jia-gu-wen; shell bone writing). They are very similar to the Shanglyphs from Shangala. It is likely they are related. At the fin de siècle of the 19th century, around the time when the Boxer Rebellion was brewing in Beijing (北京; north capital), Chinese scholars discovered that apothecaries in the capital were selling *Dragon Bone* powder ground from oxen scapulae (shoulder blades) and tortoise plastrons (abdominal shells) which had strange glyphs carved on them.

A few learned men recognized these glyphs as very ancient Chinese writings. They traced the source of the *Dragon Bones* to an ancient city called Anyang (安阳; peaceful sun) in the province of He'nan (河南; river south). Further study confirmed that nearby was Yinxu (殷墟), the Ruins of Yin, the ancient capital of the Shang dynasty.

Scholars started deciphering the glyphs and found names of Shang kings previously existing only on books written many generations later. Before this, Shang was only a myth. Unlike the decipherment of Sumerian cuneiforms or Egyptian hieroglyphs, the early efforts to decipher the Oracle Bone script did not require the presence of parallel passages of another known language, such as with the Behistun inscriptions and the Rosetta stone.

"This is fascinating," said Victoria. "I wish you or my parents had prepared me for this. I mean, I love archaeology and etymology, but ancient Chinese writing is way over my head."

"You have been in hiding and I wasn't supposed to contact your parents," David said. "In fact, your parents haven't seen me in person since they adopted you. The people who are looking for you are very sophisticated. We're supposed to hide you until you're at least sixteen, but we did not succeed. No matter, we have to move up our schedule. With your highly developed intelligence, I'm confident you'll find everything I say easy to understand."

"I'm basically a blank sheet on China," said Victoria. "If everything is a lie, how would I know you're not just making things up?"

"That is why we'll have to go on this odyssey in search for the truth," said David. "Truth is like rare treasure buried deep in the ground. It takes a long journey, a lot of time, and much hard work to discover. Shall we start?"

David went on to tell the epic story of China's civilization. The Chinese written language is the only extant writing that has an unbroken lineage connecting the modern script with its ancient progenitor, the Oracle Bone script. Most of these ancient glyphs are pictograms, and many modern Chinese words can be matched with the original Oracle Bone glyphs by sight alone. Pictograms also allow people who speak different languages to understand each other by writing. A field still looks like a field whether you call it field or call it *tian* (田). No matter what you call it, a rose is still a rose and a kiss is still a kiss. Even today, a Chinese person can communicate in writing with a Japanese person who writes Kanji, which means 'Han-word' (汉字; Han-zi).

"Chinese writing should more appropriately be called Shang word," said David. "The Shang invented the writing, the future generations developed it. Now everyone calls Chinese people Han (Simp: 汉; Trad: 漢), and calls their writing Han-word, while the true identity of their ancestor is buried and forgotten."

Who are the Han people? Why do the Chinese forget their true ancestor, the Shang, who gave them their writing in the first place? Everyone says that 90% of Chinese is ethnic Han. Few people know what Han means.

Han is the name of a dynasty, which succeeded the Qin dynasty. Qin is famous for its Emperor Qin Shi Huang (秦始皇; first brilliant king of Qin), whose tomb near Xi'an (西安; western peace) is surrounded by armies of life-sized terra cotta soldiers. Qin is also important because under Emperor Qin Shi Huang, it conquered the fractious warring states and unified China. The Qin dynasty, which was to last ten thousand generations, could not however solve the problem of succession. The empire collapsed shortly after the death of its first emperor. It lasted only fifteen years. The idea of a unified China however lives on to this day.

The name of Han as a dynasty came from the title of a commoner rebel by the name of Liu Bang (刘邦). He gained the Lordship of Han after he won the race to enter the Qin capital. The Han here means the Han River basin. It is a remote mountainous area west of the central plains of China. Kings used to exile criminals there. It was a miserable hellhole like Newcastle of Australia under the British Empire.

The Oracle Bone script of the word Han is the drawing of a man being strung up and burned as an offering to the gods (OB: ). The word means searing heat (熯: han) and is related to drought (暵: han). Liu Bang received the Han lordship as an insult and as a kind of banishment from involvement in the politics of China. He was originally promised the rich areas that belonged to the home state of Qin.

When Liu Bang finally united the country to become the emperor of China, his reign became known as the Han dynasty because he was the Lord of Han. There was no Han tribe, no Han genetics, and no Han ethnicity. Han was an undesirable mountainous region in China way out in the boondocks. The whole idea of Han as an ethnicity was created much later, which doesn't mean that it is not real, only that we should understand how it came about.

Contrary to what most people think, foreign tribes with foreign cultures and foreign genetics have frequently visited, invaded, occupied, conquered, and ruled China through the ages. Chinese history has detail records of them. China however has something that is unconquerable—its written word and the culture that is born from it. China although conquered, over time always becomes the conqueror, not by its sword, but by its word.

Most people in the West do not know, for example, that the last dynasty of China, the Qing (清; clear) dynasty, was not Han Chinese. The Qing dynasty was established by the Manchu (满族; man-zu; Man tribe), a non-Chinese tribe which has its own spoken and written language, with a distinctly non-Chinese cultural background. Their ancestors were the Jurchens (女真: nü-zhen) who came from southern Siberia and who had ruled northern China as the Jin (金; gold) dynasty during the 12th and 13th century.

By the time the Republic of China (ROC) replaced the Qing dynasty in 1912, this foreign Manchu tribe had ruled all of China for 268 years. The pigtail we see in old Western caricatures of Chinese men was a Manchurian hairstyle, imposed on pain of death on the Han population of China. The traditional long dress worn by Chinese women known as *cheongsam* (长衫: chang-shan; long dress) or *qipao* (旗袍; banner robe) also has Manchurian roots. It is now Chinese fashion. The Manchus have been born, bred, and assimilated in China for generations. To them, China is their home, and their culture is now a part of the Chinese culture. The Manchu changed China but China also changed the Manchu. They became Chinese.

"Silly me," Victoria confessed, "I've always thought that the Qing dynasty was Chinese."

"And who says you're wrong? The Manchu people are now Chinese. It's only a bit more complicated than what you can learn from the mainstream media," said David. "For people who would make the effort to dig deeper, the universe is so much more interesting. It takes the BS out of bias. You can practically scratch a Chinese and find an alien."

Plastron

Scapula

CHAPTER 26

Blessed are the Meek

In order to help Victoria understand the Oracle Bone glyphs, David told her the story of the Shang, the progenitor of the Chinaese civilization. While Victoria kept a healthy dose of skepticism in her heart, she also listened with an open mind, retaining as much as possible and taking mental notes, so that she could analyze David's stories critically and look for inconsistencies and discrepancies.

The Shang was a peripatetic tribe. It came into existence around 2000 BCE or even earlier. There are myths about the origin of the tribe, but none can explain how it got the name Shang or what it meant. The Shang was a highly civilized agricultural people famous for their religious fervour and endless wars. It became dominant at the central plains of China from around 1600 BCE until the Zhou tribe toppled Shang rule in 1046 BCE. Archaeologically speaking, the Shang exploded into the scene with a developed writing system, horse drawn chariots, and advanced bronze technologies. It was as if they had received the gift of knowledge from a Promethean god.

The Shang people were also China's first astronomers. They developed the Chinese calendar using the cycle of sixty, also known as the sexagenarian cycle, first used by the Sumerians. Most amazingly, the Shang observed the twelve-year cycle of Jupiter and used the sexagenarian cycle to count the days of the Jovian year.

"The Shang believed in many gods and spirits, and they performed religious rites at a frequency of once every other day," said David. "The Shang ruling class was particularly religious. They wouldn't do anything without first asking the gods for direction and approval. Interestingly, the Shang also believed in a supreme creator god of the universe known as Di (帝), the same word that was later borrowed to mean emperor."

"I remember this from the Yellow Emperor," said Victoria. "He is Huang Di."

"Correctamundo!" said David. "This creator god Di was often called Shang Di (上帝; up Di), meaning 'Di Above.'"

This Di name of the Shang's supreme god coincidentally is cognate with the Indo-European root *dyeu* from which we get Zeus of the Greeks and Jupiter of the Romans. Dionysus, the famous wine god, and the Dioscuri, the twin gods Castor and Pollux, are derived from the same root. Today, Italians call God Dio and the Spanish Dios. It's Dieu in French, Dev in Hindi, Dia in Irish, Duw in Welsh, etc. This is not to say that the Shang was Indo-European. It could be pure coincidence, or maybe a Promethean father of civilization had somehow taken this name of god and spread it around the world. Or maybe all human civilizations did arise from a common fountainhead.

"I have something to confess," said David. "For reasons I cannot fathom, my father decided to name me Huang Di. My *di* (弟) is of course not 'emperor' but 'little brother.' Thanks to my father, I have had to endure a childhood of endless mockery and bullying because of my name. That is one of the reasons why I decided to stay in Canada. Quite frankly, I prefer David, the beloved."

The Oracle Bone script went through several stages of development and evolution before becoming the modern Chinese script. The first development was the Bronze script (金文; jin-wen;

gold or metal writing) found on bronze vessels of late Shang and early Zhou. The Zhou later modified the Shang writing into the Seal script (篆书; zhuan-shu; carved writing), which was further modified and standardized by the Qin administration into the Clerical script (隶书; li-shu; clerk writing). By the beginning of the Tang (唐) Dynasty, scribes formalized the Regular script (楷书; kai-shu; template writing). It remains unchanged over more than a thousand years to this day as the Traditional Chinese script.

"After the establishment of the People's Republic of China (PRC) in 1949," David said, "the Chinese government wanted to make education and knowledge more accessible to the 80% of the population who were illiterate. They standardized the cursive demotic scripts, which had been in use since Han dynasty, and added new simplification ideas. These became the Simplified Chinese script."

While conservative scholars objected to simplification, practical advantages won the day. Besides, many Traditional Chinese words are themselves simplifications. In the new China under the CCP, information and knowledge are made available to everyone, not just the elites and the literati, and not just to the male gender. Knowledge is power, and it is shared with everyone who wants to learn. Simplification makes the language easier for the masses. It is a starting point to even the playing field. At least the peasants will have a chance to catch up in a very unfair race. It also amalgamates the million little villages of China into a unified nation.

Today, China's literacy is close to a hundred percent for both male and female, especially for the younger generation. The Japanese borrowed it and called it Kanji. The Koreans used it before the 20th century; they called it Hanja. The Vietnamese used it before the French colonized their country, and they called it Han-van. No one remembers to give credit to the Shang who invented the Chinese writing in the first place.

"Chinese is actually easy to learn," said David.

"I am glad to hear that," Victoria exclaimed. "I figure that I'll have to learn some Chinese in order to talk to my real parents."

Here are the reasons why Chinese is easy to learn. The Chinese writing is monosyllabic when used with the Chinese spoken language. It has no conjugates, it is gender neutral, it has no singular or plural, it has no past, present, future, or continuous tense, and there are no new words. One only needs to learn about a thousand words to survive in modern Chinese society. Usually, a couple of existing words can be put together to mean something new. For example, computer is 'electric brain' (电脑; dian-nao), and television is 'electric sight' (电视; dian-shi). The word 'electric' (电; dian) itself comes from 'lightning.'

"The unique Chinese picture writing is in fact a very good global language, maybe even a galactic language. It allows different tribes sharing a vast land but speaking different languages to communicate," said David. "It does not matter that for a large part of Chinese history, the country was under the rule of tribes and cultures that were foreign to the native inhabitants. All thanks to the Shang word, the conquerors would eventually become a part of the Chinese civilization, a civilization as old as the mountains and as deep as the sea, and it's a sea that accepts a hundred rivers."

Shang had a very strong military, but the dynasty was overturned by rebellion and betrayal. Shang culture however lives on in all Chinese because of its unconquered writing. Those who worship violence cannot understand the virtues of cultural power. A most relevant example is that of the Jewish people.

The kingdom of Judah lost its royal house and its holy temple after the fall of Jerusalem to the Babylonian Empire in 586 BCE. When future Roman emperor Titus sacked Jerusalem and destroyed the second temple in 70 CE, a large number of Jews went into diaspora to the four corners of the earth, without a temple, without a country, without a leader, certainly without an army, and they were always subjected to discrimination and persecution. Yet the Jews outlived the Babylonian Empire, the Roman Empire, the Byzantine Empire, the Mongolian Empire, the British Empire, and every other empire you can think of.

"Two thousand and six hundred years is a long time," said David. "All the Jewish people had was a book and their determination to survive."

Why can't people see that cultural power has much better survivability than military power? The ancient sages have been teaching this since antiquity, only the people have not listened. People never do. They can't hear anything when firing their guns.

Lao Tzu said, "The softest and weakest will overcome the hardest and strongest." In The Sermon on the Mount (Matthews 5:5), Jesus said, "Blessed are the meek; for they shall inherit the earth." The Sage Yellow Stone (黄石公; huang-shi-gong) was quoted in history as saying, "The soft can conquer the hard; the weak can conquer the strong."

"People who believe only in the virtues of violence and war are sadly mistaken," said David. "Those who think superior military is the only measure of success will be swallowed by time like all the others before them."

David began to wax poetic on the fall of empires. "My name is Ozymandias, king of kings; look on my works, ye mighty, and despair!"

Victoria, who knew Percy Bysshe Shelley's *Ozymandias* by heart, joined David in harmonious recitation. "Nothing beside remains, round the decay of that colossal wreck. Boundless and bare, the lone and level sands stretch far away."

Chapter 27

The Oblique Order

Before the battle between Queen Zia's Shang army and the barbarian Gotts began, it would be appropriate for battlefield observers to understand the symbols on the banners of the Shang army.

The most numerous flag in the Shang army was of course the Shang flag. The Oracle Bone symbol for Shang is ( ) with quite a few variations. It changed over time into the modern form (商). As a pictogram, the symbol looks like a structure for making offerings or sacrifices to the gods. The top part is a dagger (辛: xin; later means bitter; OB:) used for ritual slaughter, but sometimes also for tattooing a concubine or blinding a slave. The lower part is the altar with storage space underneath, sometimes with a receptacle for catching the blood of the sacrificial victims. A variation has stars to indicate that the word is the name of the Shang Star ( ), the national star of Shang, also known as the *Heart of the Dragon.*

Queen Zia's (妇好; fu-hao) banner shows ( ), the symbol of a female (女: nü; OB:) with a male child (子: zi; OB:). This composite word translates to the modern Chinese word that means 'good' (好: hao). Her clan name is however not *Hao* (好: good), but *Zi* (子: son). It was customary in ancient times to add the female radical to a name to indicate the name belonged to a female or a matriarchal clan. *Zi* (子; son) happens to be the clan name of the Shang's royal house. That is how we know Queen Zia (妇好;

fu-hao) was King Wotan's (武丁; wu-ding) blood kin. Queen Zia's title *Fu* (妇; ), which is the drawing of a broom, in modern Chinese means 'married woman,' but for Fu Hao, it denotes the wife of the king and therefore meant 'queen.' Fu Hao is therefore Queen Hao, more accurately Queen Zi, known to us as Queen Zia.

Lady Diane's banner showed the symbol for 'field' (田: tian; OB: ). The same symbol in Bronze script (BR: ) or one showing a man next to his field is the title of a lord known as *Dian* (甸; BR: ). The word for 'field' (田: tian) in ancient times also meant 'hunt.' Lady Diane did not add the female radical to her symbol because she was not a wife, and she wasn't going to start a matriarchal clan. Like the Greek goddess Diana, she would remain a virgin huntress.

The queen's army flew King Wotan's flag to create the illusion that the martial king was present. His Chinese name is Wu Ding (武丁; martial nail). The word *Wu*, which means 'martial' shows a halberd and a foot (武; OB: ). The word *ding* (丁; nail, OB: ) showing the head of a nail is actually Shang's counting system based on the *Ten Celestial Stems*. *Ding* is the fourth position. Wu Ding anglicized would be King Martial IV. Wu Ding is famous as one of the greatest kings of Shang, and in this tale, he is known as King Wotan.

Finally, the Gotts was a barbaric tribe living to the northwest of the Shang. Contrary to the rumours of the times, the Gotts were highly civilized and were in constant contact with people from the south. They also acquired the technology to produce weapons, tools, and vessels of bronze. The Gotts would prove to be a much more difficult foe than most of Shang's neighbouring tribes.

The Gotts worshipped the spirit of the wolves. When they went to war, their Chief's personal guards carried a large spear adorned with wolf tails and wolf fangs. Their crack troops wore a fierce wolf mask to terrify their enemies. The Shang called them by a name that meant 'ghost, god, or spirit' (鬼; gui). The Oracle Bone script () shows a man wearing a mask. The Shang was generally terrified of the Gotts.

We now return to the battlefield. As Lady Diane had ordered, the Shang lines were drawn up even before the first rays of the sun peeked over the horizon from the southeast, with the centre division under Queen Zia accompanied by Lady Diane, and the wings led by the queen's left and right wing generals. The autumn morning felt nippy, the sky was clear, and there was no biting wind from the north; hence, the weather was perfect for slaughter. The queen and her trusted field marshal looked at each other and nodded to acknowledge that the game was afoot. The queen gave the order for the battle drums to roll, signifying that it's time to rumble, and soon the whole line marched steadily forward.

This was not a large-scale engagement by any stretch of the imagination, but it involved all of the queen's own citizen soldiers of about three thousand and another one thousand armed men belonging to Lady Diane. There was about five hundred support staff consisting mainly of servants, slaves, and camp guards who could fight if they had to.

The Shang army had fifty chariots, each supported by fifty hoplites. Each chariot had a charioteer, a phalanx commander who was an expert archer fighting on the left side of the chariot, and a man-at-arms wielding a Shang halberd on the right. The chariot was used as a mobile command vehicle for the phalanx, not as a shock weapon. The phalanxes and the chariots were required to advance orderly to the beat of the war drums.

The Gotts were surprised that the Shang lines were already marching towards them without the usual rites and ceremonies. There was no time for eating breakfast. They had to quickly arm themselves and form a defence line.

Lady Diane gave the order for the first action of the day. A small detachment of light mounted skirmishers at the left wing galloped off to probe the enemy's right flank. The Shang had never used this kind of tactics before. The skirmishers were ordered not to engage but to fire their arrows as they came within range, and then reel to their left to safety and regroup to make another charge. Since the Gotts were using less effective bows and arrows of shorter range than the Shang's composite bows, the Shang skirmishers could practically harass the enemy all the livelong day without suffering any losses. Lady Diane had invented a saddle by which the mounted archers could ride securely fasten to their horses, so that they could turn around and fire arrows at their pursuers without falling.

The horde's chief, who commanded a squadron of heavy cavalry at the centre, led his armoured horsemen to the right wing to counter the Shang threats. He believed that most of the heavy fighting would be at his right wing because his left wing was flush against the riverbank, which was swampy and not easy for either the Shang chariots or their hoplite phalanxes to move over. The right wing was also the only possible escape route for the Gotts.

Lady Diane then gave the next order for a difficult maneuver. The left wing of the Shang army would stop marching, forming the phalanxes into interlocking defensive squares in anticipation of the Gotts' cavalry charges. Queen Zia's central division would march and pivot anticlockwise so that its right flank was closer to the enemy lines. The Shang right wing then made a direct charge at the lightly defended left wing of the Gotts.

Lady Diane developed the oblique order to attack the enemy's weaker wing while denying the enemy's stronger wing. It was an idea far in advance of its time and was lost for six hundred years until its employment at the famous Battle of <u>Chengpu</u> (城濮).

The oblique order was then used by the Theban general Epaminodas at the Battle of Leutra against the Spartans, followed by the audacious Alexander at the decisive Battle of Arbela (or Gaugamela), defeating the much larger army of Darius III of the Persian Empire. Almost three thousand years after Lady Diane, Prussian king Frederick the Great would use this battle order to great effect at the battles of Rossbach and Leuthen, defeating enemy armies twice as large. Victoria also used it regularly when she employed Sicilian Defence while playing chess with Michael.

As the struggle at the Shang right wing was hanging in the balance, Lady Diane made an order to send out the reserve troops. A squadron of heavy cavalry using long lances rode out from behind the hill and charged the left flank of the Gotts.

Lady Diane developed the Lancers reserve with the idea that the best way to counter the Gotts' cavalry was to use cavalry. This was a big surprise for the Gotts, as everyone knew that the Shang army favoured the plodding tactics of the orderly phalanx and chariot. The Shang wasn't supposed to have a formal cavalry. The armoured Shang horsemen secured to their saddled horses, wielded a long lance to cut and stab the Gotts' horsemen at a distance. Lady Diane developed the lance to gain an advantage over the more commonly used spear.

As the lancers rode out from behind the grassy knoll, the Gotts' heavy cavalry at their right wing launched a counter-attack. They dispersed the Shang skirmishers and charged the Shang left wing. The masked Gotts looked like fiery angels of death riding on monstrous beasts bringing apocalypse to the shaking Shang hoplites. The queen in the centre now brought forward her archery squadron and from their slanted forward position, fired arrows at the charging Gotts, forming a lethal crossfire with the archers of the Shang left wing, which effectively blunted the Gotts' attack.

By the time the Gotts' heavy cavalry arrived at the Shang left wing, the momentum of the charge had dissipated and the attack became disorganized. The Gotts were not able to break any of the well-defended squares. When the Gotts chief saw that his heavy cavalry was incapable of causing damage to the Shang left wing, and that the Shang skirmishers were regrouping for a counter-attack from the flank, he ordered a retreat.

The armies had been on the field for several hours. The shadow was getting short, and the Gotts were getting hungry from having missed breakfast. As the Gotts chief was riding back to his lines thinking about what to do next, he received the news that the Shang had broken through his left wing. He immediately dispatched troops from the centre to stabilize the broken wing before it turned into a rout.

The chief realized that he might have met his match. He was left with only one thing to do, and that was to charge the queen's elite troops at the centre. The Gotts chief rallied his horde for one desperate charge—the Hail Mary play.

"Wolf brothers," thundered the chief as he rode up and the down the front of his rallied troops, "the Shang cowards are giving us their women so we can have beautiful wives. The brave horses at the front get the women; the slow horses at the back get the goats."

The horde roared with laughter. The chief exhorted his troops, "What are we waiting for? Let's go get 'em."

Without waiting for the chief, the horde rushed forth like a hurricane towards the queen's oblique line in the centre, howling like the spirits of the wolves in the mountain range of the seventy black peaks, better known as the *Mountain of the Great Canopy.*

CHAPTER 28

Twilight of the Gotts

As the Lancers rode into action on the Shang right wing, Lady Diane saw that the battle was arriving at a critical juncture. She turned to the queen beside her to make a special request.

"Your Majesty," Lady Diane said, "I ask for permission to take the field. It is important that I personally lead the Lancers to close the ring on the Gotts. All the centre needs to do is to keep the phalanxes fixed in their positions and order our archers to fire volleys into the enemy. Our engineers had buried empty wine amphorae in ditches last night to cripple the Gotts' charge. I have just sent orders for the left wing to assume attack formation and to envelop the Gotts. We will not let a single barbarian escape today."

"Can't you send someone?" Queen Zia said, "I don't want you to take unnecessary risks."

"Your Majesty," said Lady Diane. "I need to go now or the Gotts will escape. Wish us luck. I will return anon with good news."

Lady Diane kissed the archer-ring on her thumb as a public display of her friendship with the queen, and then leapt off her chariot without waiting for the queen to respond. She mounted on the shining black steed that was waiting for her, grabbed a long lance from her squire, and commanded the captain of the Royal Guards to keep the queen safe until her return.

Lady Diane's steed reared in the air and with a mighty kick shot like an arrow into the distance. All eyes friend and foe alike marvelled at the sight of this shimmering black beast, thundering across the battlefield like a monster from Tartarus, while Lady Diane's black cape billowed like the waves of the roiling Yellow River that cascaded down the cataracts of Dragon's Gate. Lady Diane traded for this black steed from a caravan that came from the land beyond the setting sun. His name was Swarh Li (什伐骊: shi-fa-li), which combined the foreign word Swarh (ride) with the Chinese word Li (骊; black horse) to mean Beautiful Black Steed.

The Shang's right wing with the help of the Lancers broke the Gotts' left wing. The battle had gone according to plan so far. Lady Diane was however worried as she saw an ominous cloud of dust rising behind the horde's daring charge at the queen. There was not a moment to waste. She would lead the Lancers on a flank charge at the Gotts and deliver the coup de grâce.

It was a sad ending for the Gotts. Their fastest horses did not get the pretty maidens. Their iron hooves crushed the buried amphorae and fell into the ditches prepared by Lady Diane's engineers. Then a torrent of arrows rained on the unlucky Gotts that came near the queen's phalanxes, sending the eager riders on a fast track to Valhalla with the Valkyries as their new brides.

Almost at the same time, the Lancers led by Lady Diane arrived and smashed into the left flank of the charging Gotts. The carnage was horrific. The Gotts' desperate charge failed. They were trapped in a pocket with retreat the only route. The Gotts chief could only try to escape with his life and avenge this insult at another time and place. With just a hundred of his best fighters, he fought his way back to the wagon fortress, hoping to rally the troops and find some ways to escape.

The Gotts chief fell into despair when he found that the Shang had already taken the wagon fortress. Lady Diane had sent out her

Special Operations platoons in the dark of the night to make a wide westward flanking march through the woods to the marshes behind the Gotts' line, and there waited in hiding until the moment was ripe for them to take the wagon fortress by surprise.

Leading a unit of lancers, Lady Diane and her personal guards cut a swathe through the Gotts and arrived at the Queen's front. Without dismounting, Lady Diane saluted to the queen and delivered the news.

"Your Majesty, I bear great tidings," she said. "We have won a glorious victory today. We have surrounded the Gotts on all sides. Please order your chariot phalanxes to march forward without leaving any gaps. The ditches in front have been prepared with marching lanes. I will now capture their chief and bring him before Your Majesty."

"After the battle," said the queen, "I will deal with you for leaving against my will."

Queen Zia sounded stern, but she secretly stole a subtle smile.

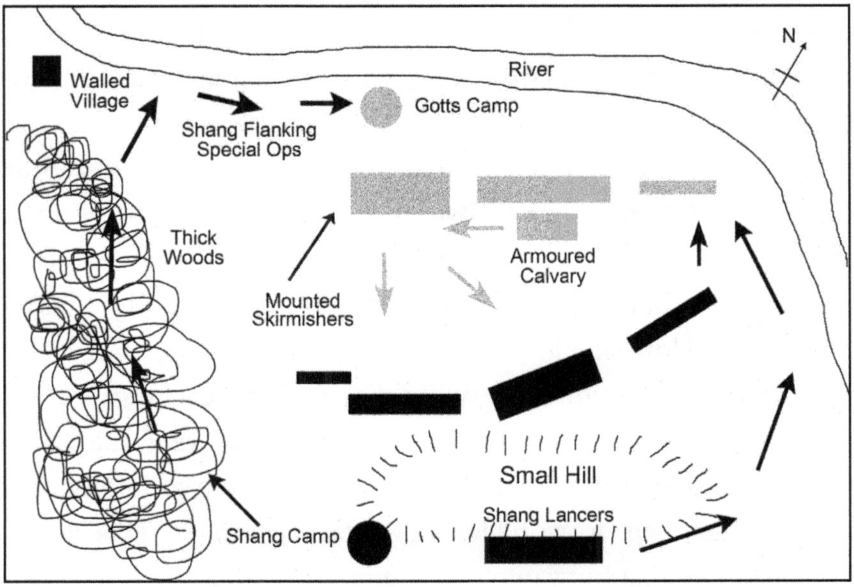

"As you wish, Your Majesty," said Lady Diane. She knew that this complete victory over the Gotts would absolve her of any mischief. The battle was practically over, but Lady Diane wanted to make the victory more meaningful than just another day of wanton slaughter. Lady Diane was born to be a warrior, but she had always yearned for peace.

"I have a request," she said. "The Gotts are a valiant and clever people. They are more civilized than the fable-mongers have painted them. Rather than slaughtering them wholesale, I believe it may be to our advantage to have them as subjects or even allies."

"They have caused much death and suffering on our side," said Queen Zia. "Do you think the king will not want to extract some satisfaction of vengeance from this victory?"

"As Your Majesty knows, there is no profit in revenge. We should consider actions that will bring long term benefits to us," Lady Diane said. "The universe is most benevolent when it is in balance. We have achieved so complete a victory today, that we must ourselves restore the balance. It is when we have acquired great power that we should show great mercy. For the sake of the kingdom, I ask that the Gotts be spared provided they submit to the King and become subjects and allies of Shang. If they agree, then the kingdom will have a faithful and strong ally to help keep the peace for our borders. Don't we want our children to grow up in peace?"

"What is our excuse for this radical peacemaking?" asked Queen Zia. "How should I explain to the king?"

"This is in accordance with the ancient teachings of our ancestors," said Lady Diane. "In the spring, we're forbidden to hunt and slaughter the pregnant and the young. We're forbidden to cut saplings and to conduct aggressive warfare. This is the understanding that by wanton slaughter, we will destroy ourselves. We're just following the ways of our great fathers."

“Do your best then,” said Queen Zia, “but be careful.”

The chief of the Gotts by this time had only a handful of men left with him. They were hungry, thirsty, and tired. The chief had tried but failed to break out of the ring of Shang hoplites. The only thing left for him to do was to fall on his own sword and end this humiliation. But even in this, he was not successful. His faithful steed walked into a trap and the Shang soldiers promptly netted the chief alive, along with about two thousand Gotts who survived the battle.

Lady Diane arrived at the prisoner’s camp with her mounted guards. She rode up to the prisoners to have a good look at their condition. They were a pitiful sight. Almost everyone had wounds. Yet they stood proud in front of the woman who thrashed them on the battlefield.

“I am Lady Diane, chief counsellor and field marshal for Queen Zia of the Shang King Wotan,” yelled the hoarse female fighter. “I wish to parley with your chief.”

Not a single prisoner moved or uttered a word. A few Gotts in the front spat in contempt. One of the Gotts with half of his face covered in blood started laughing, spitting out more blood as he did so. Another one with an arrow through his shoulder and a big gash on his face followed suit. A third with his cracked skull in a bloody bandage was literally in stitches. Soon the whole pathetic group was laughing uncontrollably, half of them holding their stomachs and the other half with tears rolling down their face. It was impossible for the Shang to understand the humour of the Gotts, but they were men who would laugh in the face of death.

Lady Diane made a gesture with her hand towards the pile of firewood that was collected for the camp’s frigid night. It instantly burst into flames. The Gotts were no less superstitious than the Shang. This woman in front of them must be a powerful witch. That was enough to shut them up.

"Do you know Queen Zia had fought ten great battles against neighbouring tribes and won every one of them?" shouted Lady Diane. "Have you not heard about our famous battle with the Sand People, how we levelled their walled city and captured the flower of their youths? Do you know what happened to the captives? Half of them, the proudest and best fighters, became our slaves; many of them became eunuchs, the other half we sacrificed to the gods."

A murmur spread across the prisoners like a ripple in a pond, especially at the word 'eunuch.'

"I do not want such a bloodbath," Lady Diane said, "I want to make an offer to your chief. If the Gotts make an oath of friendship with the Shang, we will set you free to return home to your wives and children, with your horses, with food and gifts from the Shang King Wotan and Queen Zia. The other choice is the fate of the Sand People. So where is your chief?"

The Gotts chief slowly sauntered up to Lady Diane's black steed, and caressed the top of the animal's nose.

"We are the Gotts," the chief declared. "We have lost today to the connivance of a whore—a witch who practises the black arts and sleeps with the farting devil (a few laughed), an old hag whose foul stench repels all men. We would rather die than submit ourselves to you. If the Shang wants the Gotts to be friends, let your bravest fight me in single combat to the death. If I die, the Gotts will be your subjects. If I live, then we go home with our plunder."

"Good, I challenge you," said Lady Diane. "Guard, give the chief a horse and a weapon of his choice."

"I will not fight a stinking strumpet," cried the chief with contempt as he spat on the ground.

"You cannot choose," said Lady Diane. She rode over and poked the chief with the butt of her lance. "I have made the challenge, unless you're afraid of losing to a wench."

"In that case," cried the chief, "harlot, prepare to die."

The chief grabbed a spear from a guard and leapt on a horse that was brought over for him. Let the clash begin.

The warrior chief was however no match against Lady Diane, who was an expert fencer with the lance. After toying around with the chief, Lady Diane disengaged from the tussle and galloped off. But the chief thought that he had the upper hand and that his feminine opponent couldn't handle the physical exertion.

As the Gotts chief gave chase, Lady Diane suddenly twisted her body around to make a backward thrust with her lance. The chief hurriedly parried but Lady Diane avoided the chief's spear with a circular counter-parry, went under his arm, and whacked his side with the pole of her lance, careful not to cut the man. It happened so quickly and so unexpectedly that the chief was knocked clean off his horse. Just then, Queen Zia arrived on her chariot followed by her Royal Guards and her pack of wolfdogs.

"Lady Diane; we owe much of today's glorious victory to you," said the queen. "Therefore I decree that your reward will be the authority to deal with the prisoners as you please. The sun is setting. We have pledged our victory to the Victorious Sun and the Lord God Di. The offering will begin soon. We need you to be our chief priestess. The gods will be overjoyed from having so many men sacrificed to them. Otherwise, we have a large number of sacrificial animals, including three hundred white oxen. Settle this quickly and don't be late." With that, Queen Zia turned her chariot around and left the scene.

The chief was still lying in the dust, a bit dazed from the fall. At a time like this, a fall was the necessary dose of reality for the Gotts. If the Shang had such a woman to lead them, the Gotts didn't stand much of a chance. Perhaps the Gotts could learn from this woman general and be stronger for it, and not just in the military sense.

Lady Diane dismounted and offered the butt of her lance to the chief. He grabbed hold of it and pulled himself up.

"Well, now that we have fought," Lady Diane said most magnanimously, "will the Gotts befriend the Shang?"

The chief returned the gesture by raising the lance with Lady Diane's arm and the prisoners roared a shout of approval.

Even though Lady Diane's voice was hoarse from a day of shouting, she needed to make her last speech of the day.

"Brave warriors, tonight you will rest in safety," she shouted, "you will have hot food, and our shamans will attend to your wounds. Tomorrow, we will invite your chief and your chieftains to represent all the Gotts at the Ceremony of Brotherhood, when we will make the oath of blood to the gods. Afterwards, everyone will wine and dine together and share food as family. The past quarrels will be forgotten."

Lady Diane made a gesture to her squire, who quickly handed over a rolled up banner. She unfurled it to show the sign of a scorpion.

"This is the banner you will use while you are in the Shang camps," said Lady Diane. "We call it the flag of the Tuman, the chief of the host. The royal court of Shang will honour your chief from now on by this title. Our soldiers will fight side by side against common enemies. There will be no more raiding of Shang villages. We will teach you how to trade instead. No one needs to die and everyone will profit. What say you?"

The chief raised his Tuman banner high up in the air, right beside the banner of Lady Diane and the Shang flag. "Tuman, Tuman, Tuman," chanted the Gotts.

Chapter 29

Children of the Flower

Victoria guessed from her *huang* jade pendants that her family must be secretly connected to the royal lineage of Shang. Even Confucius did not advertise his royal Shang bloodline. Secrecy was perhaps the Shang's best survival policy. David however made it clear that Victoria's background was a lot more complicated than that.

"I know it's a long and winding road," said David. "But by the time you learn your true identity, it will be less of a shock to you."

He began to give Victoria a concise and abridged historical account of the various alien tribes that had contributed to China's genetics and culture.

During the *Spring and Autumn Era* of the Zhou dynasty, which followed Shang, a foreign tribe established a vassal state known as the Zhong Shan (中山; central mountain) state in northeastern China. Historians called this tribe Baidi (白狄), which literally means 'white northern barbarians.'

When Qin unified China in the third century BCE, it established a unified Chinese culture that resisted fragmentation and conquests by foreign tribes. Few remembered that Qin was itself a foreign tribe from the west.

During the last years of Qin, a tribe known as Xiongnu (匈奴; dialect: Hun-nu) became powerful in the north. The Han Empire fought endless wars against Xiongnu and drove the northern half of the group into Central Asia, which later migrated westward into

Europe as the Huns. The Southern Xiongnu moved south to live among the Han, intermarried, and became assimilated. Life was a lot easier. The food was hot, the beds were soft, and the women were pretty. Resistance was futile.

Before the Han dynasty fell apart in the third century CE, a large complex barbarian tribe known as Xianbei (鲜卑) moved south into China from the northeast, eventually controlling northern China under the rule of what Chinese historians called the Northern Dynasties. The cradle of Chinese civilization was under foreign rule for two hundred years.

"In truth, northern China for more than three hundred years," said David, "was overrun by all of its neighbouring barbarian tribes."

"I'm having a mental picture of Costa Rica, Panama, and El Salvadore overrunning America," exclaimed Victoria. "What happened to mighty China?"

"China when divided is easily conquered," said David, "but the Chinese culture is invincible, and it eventually conquered the barbarians. A very good example would be the Xianbei Northern Wei (北魏) dynasty. One of their Xianbei kings made laws banning the usage of their own language, forcing their people to speak Chinese, adopt Chinese names, practise Chinese customs, and intermarry with the Chinese. They wanted to become Chinese. Eventually, Xianbei rule was usurped by the short-lived Sui (隋) dynasty, which was then succeeded by the glorious Tang (唐) dynasty."

"Should we take a break?" Victoria asked with concern when she saw David's face turning pale, probably from overexertion. He was after all middle-aged and it had been a long and hectic day.

"I'll be fine," said David, stoically swallowing some pain that was obviously bothering him. "I'd like to talk about Tang dynasty. It is an important part of your story."

Tang was a very interesting and influential dynasty. Its capital Chang'an (长安; long peace; modern Xi'an) was the greatest city in the world at the time. Christians, Moslems, and Buddhists from other parts of the world came to spread their religions and to trade with China. Even today, Chinatown in Chinese means 'Tang Person Street' (唐人街; tang-ren-jie).

The founding emperor of Tang dynasty Tang Gaozu (唐高祖; High Ancestor of Tang), whose mother was a Xianbei princess and whose father was probably a sinicized Xianbei, married a wife of predominantly Xianbei lineage, making his son, the great Emperor Tang Taizong (唐太宗; Great Patrician Father of Tang) more Xianbei than Han-Chinese. Tang Taizong's queen was again Xianbei, making the next emperor predominantly Xianbei. One can imagine there was a lot of Xianbei influence in the presumably Han-Chinese Tang dynasty.

The Xianbei's sinification was not a one-way street. The Xianbei also changed the northern Chinese. For example, it was Xianbei custom for the woman to run the household. Many northern women were well educated and strong-willed, allowing them to gain power and status in society. A good example of a northern Chinese girl from the Xianbei period would be Hua Mulan (花木兰; Flower Magnolia) made famous in the West by Disney.

Mulan's family name *Hua* (花) is the Chinese word for 'flower.' It is related to the more popular family name *Hua* (Simp: 华; Trad: 華; OB: ) which means 'elegant' later having the meaning of China. In ancient times, hua-flower and Hua-China originated from the same symbol—the flower bush. A Chinese (华人; hua-ren) is an 'elegant person' or a 'flower person.'

"Hua Mulan was a fabrication," said David. "A real girl who was very special in Chinese history emerged in that era. She was a lowly concubine of Tang's greatest Emperor Tang Taizong. The young girl was fourteen years old, not much older than you, when she entered the harem of the forty-one year old Emperor."

"Gross!" Victoria cried.

"It is gross now, but entirely appropriate in those days," said David. "As the story goes, the young concubine was childless when the emperor died twelve years later. By tradition, she was supposed to spend the rest of her empty and meaningless life in the royal convent. If she did, we would never have heard of her. Tang Taizong's son, the next Emperor known as Tang Gaozong (唐高宗; High Patrician Father of Tang) somehow caught sight of his beautiful stepmother at the monastery, decided to rescue her, and promptly made her his concubine."

"Disgusting!" cried Victoria as she recoiled into a heap and covered her face with her hands.

"It may be disgusting for the conservative Chinese, but perfectly normal for Xianbei," David continued. "Tang Gaozong eventually made his stepmother-concubine the official queen consort, replacing the original queen. After the death of Tang Gaozong, the queen would establish her own dynasty known as Second Zhou (后周) and ruled over the entire Empire as supreme ruler. Her name is Empress Wu Zetian (武则天; Martial Engrave Heaven), the only female emperor of China."

Empress Wu was not unique in that she was following in the footsteps of strong female royalties of the Northern Dynasties, each one with a life story just as incredible, but to take that final step to establish her own dynasty was revolutionary. Furthermore, Empress Wu's rise was no accident. She was the descendant of a secret matriarchal clan of ancient Zhou. The clues are hidden in plain sight. Victoria was closely related to this clan as well.

Women during this period gained unprecedented status in China. There was no foot binding. Girls received an education. They were involved in the law and in literature. They even led armies into battle without having to disguise themselves as men. The great Tang Taizong's younger sister, Princess Pingyang (平阳; calm sun), independently raised and led an army of seventy thousand when she was nineteen years old. She fought many battles that helped to establish the Tang dynasty.

Some of Tang's liberated women worked as professional courtesans. They were similar to the Greek hetaira, which means 'sophisticated female companion.' They could speak on many subjects intelligently among highly cultured and aristocratic clients. Many of these courtesans were respectable and independently wealthy. Society did not shame girls for having premarital sex. Women were free to wear men's clothing and to go around on horseback. Tang women were no prudes. They developed a full figure and wore daring low-cut dresses that showed off their décolletage. Walking down the streets of the Tang capital Chang'an, one might mistake the city to be Gay Paree.

"Empress Wu was not just a pretty face. She was a very intelligent ruler who was constantly being challenged, including by her own children," said David. "Many capable men served her faithfully. Under Empress Wu, the empire was mostly peaceful and prosperous. She died of old age as one of the longest living emperors in the history of China at eighty-two years old. But Empress Wu could not single-handedly change the patrilineal succession scheme. Her Zhou dynasty would last for only one term. Her son ascended the throne and promptly changed the dynasty back to Tang. That's the story of the lowly concubine who became the one and only female emperor of China."

"That's an amazing and inspiring story," Victoria said. "But I need to take a break even if you don't need one. I'm having a really bad headache."

Victoria made herself comfortable lying down across the back seats.

"I've been thinking, Mr. Huang," she said, "aside from that one time you've had to protect me from a gunman, you're basically a gentleman and a scholar. You should be a professor or something."

"I was a professor," said David wistfully. "After getting my PhD at University of Toronto, I taught ancient Chinese culture at their Department of East Asian Studies, and I was hired by the Royal Ontario Museum to catalogue their large collection of Oracle Bones donated by Minister James Menzies and Rev. William White. I would've died happy working both jobs. But ROM funds dried up, I didn't get tenure at U of T. All of a sudden, I was unemployed. I had just gotten married, bought a house in Toronto with a hefty mortgage, and my first child was on the way. What could I do but to use my knowledge to sell myths? A person who depends on a salary cannot afford the truth. When people in China got rich, I found easy victims. My art and antique business became lucrative. I just never looked back."

Victoria was already asleep. She might not have heard David's confession. David took a fleece blanket from the storage bin and pulled it over the teenage girl.

Shang halberd (ge; dagger-axe 戈)

Chapter 30

Book of Secrets

When Victoria woke up, Edvard Grieg's Peer Gynt Suites was playing in the background. The music really calmed her nerves. It was Michael's favourite melody in the morning.

"Good afternoon, just in time for lunch," said David. "How's your headache?"

"Wow, I slept for a long time. I'm fine now, thanks," said Victoria with a yawn. A short blast of pine-scented air replenished the oxygen in the cabin while a robot arm offered a mint and a wet towel. "Nice touch. Where are we?"

"We just passed Medicine Hat. Calgary is coming up. We should be arriving at Field in five hours," said David. "Moira got us rooms for the night and reserved a table for us at Truffle Pigs. We'll have a proper Christmas Eve dinner."

"I don't mean to pry, but how's your family taking this?" asked Victoria, "You being not at home for Christmas."

"They think I'm at Mumbai," said David. "I'll call them later. My duty towards you is the same as my duty towards humanity. I love my family, but in order to save them, I must first keep you safe and take you to my father."

"I'm not sure I can accept everything you say so far," said Victoria, "but I hope everything works out for you and your family. By the way, what is Field?"

"Field is just a field of snow in the Rockies," said David. "But if you want to check out the wonderful life of the Burgess Shale, Field is the starting point."

"The Burgess Shale … isn't life strange!" remarked Victoria, who had read Alan Jay Gould's *Wonderful Life* about the weird Cambrian life forms of the Burgess Shale. "I'll have to come back someday when all this blows over."

"And for listening patiently to my story, Moira has asked the chef at Truffle Pigs to prepare a special pork dish for you," said David. "It's a Yan Can home recipe with mango and chestnut. We can pretend we're at home. After all, home is where the hog is."

"Is that the slogan of Truffle Pigs?" Victoria asked, "Or did you just make that up?"

"I made up the slogan, but everything I have told you is a hundred percent factual," said David. "Chinese people have a close relationship with pigs since antiquity. The Shang people kept pigs in their homes. In fact, the Shang word for 'home' (家: jia) is a pig under a roof (). Now you know why you love pork so much."

"I know," said Victoria, "I've got Shang genes."

"You'd be surprised. You may even have Tibetan genes," David said. "Do you know in the middle of the eighth century, the Tibetans captured Chang'an, the Tang capital? They probably did some gene mixing. The Tang also sent genes the other way. We mustn't forget about Tang's Princess Wencheng (文成), who became the queen of Songtsen Gampo, the founder of the Tibetan Empire. She is credited to have introduced Buddhism among other cultural elements into Tibet."

David continued to tell the story of alien tribes in China through the ages. Tang was a time when visitors from distant lands were welcome to live in China and trade with China. Many well-known officials of Tang were foreigners. The list of alien tribes living in China or interacting with China is bewildering. The earliest Turks in China were known as Göktürks (突厥: tu-jue) and Shatuo (沙陀) Turks. Both had set up significant reigns in the region. There were many famous Turks in Chinese history.

The Uyghurs (维吾尔; wei-wu-er), a complex tribe ruled at one point by the Göktürks, were known as Huigu (回鹘) in Chinese history. The nature of the tribe has changed over time but their connection with China is long and deep. Today, quite a few ethnic Uyghur Chinese actresses are very popular in China, including Gulnazar Bextiyar, Medina Memet, and Dilraba Dilmurat just to name a few. They will change people's idea of what a Chinese person should look like or what kind of name they should have, but they're Chinese just the same.

After the Tang dynasty collapsed, China once again devolved into chaos, and the foreign tribes were again in the thick of it. A Chinese general with secret Shang connections established the Song (宋) dynasty, taking its name from the Song state, which was ruled by the Shang royal house. It is an epic secret which is hidden in plain sight.

Around the same time, an iron tribe became powerful in the northeast. They were the Khitans (契丹; qi-dan), a descendant of the Xianbei. They established the Liao (辽) dynasty and ruled northern China for a couple of hundred years. The Russians thought that they were Chinese and even today call Chinese people Kitay. The English anglicized Kitay into Cathay, which means China. In time, the Khitans followed the footsteps of their Xianbei ancestors and became Chinese. One of Khitan's branches, the Shiwei (室韦), would eventually develop into the mighty Mongols.

Next in line to rule north China was the Siberian Jurchens who established the Jin (金; gold) dynasty. The Mongolians under Kublai Khan, the grandson of Genghis Khan, overthrew the Jurchens and ruled the whole of China as the Yuan (元; beginning) dynasty, until a lowest of the lowest caste Chinese wandering mendicant monk implausibly became the founding emperor of the Ming (明; bright, enlightened) dynasty.

After Ming, the Manchu descendants of the Jurchens ruled all of China up to 1912 CE. China is a big country shared by many different peoples each with their own story to tell. Some of them are only remembered in history while others are still around. Modern China after all recognizes fifty-six distinct ethnicities and their native cultures in the country.

"Then how does one," asked Victoria, "identify someone as Chinese?"

"The Confucian Golden Rule says, 'Do ***not*** unto others what you do not wish others to do unto you,' and that is the core of the Chinese culture," said David. "I would say adopting Chinese culture and respecting our roots are the criteria for being Chinese, not some specific Han gene. The rest are just empty names."

"Chinese history sounds stranger than fiction," exclaimed Victoria. "You told me that a commoner became the emperor of Han, a concubine took over the Tang, and a beggar became the emperor of Ming. Chairman Mao was a peasant's son and he became the leader of China. Is this normal? How does a peasant, a concubine, or a beggar rise from nothing into a supreme ruler? Do you think providence have a hand in these improbable events?"

"Almost every dynasty in China failed at least in part because of peasant rebellions," David said. "Only the most asinine Chinese rulers in history would ignore the plight of the peasants, usually to their own detriments. On the other hand, peasant rebellions more often than not devolved into mob rule and chaos because they lacked good leaders. But now and then, someone extraordinary would rise to the top, and unite the country in peace and harmony. Of course they received help, but not from providence."

If it is acceptable to believe that peasants, beggars, and concubines can become supreme rulers of China's greatest dynasties, then it is not farfetched to believe that extraordinary people with great abilities helped them. Chinese history recorded that the founding of the Han dynasty received help from a mysterious man known as Sage Yellow Stone. He gave a mysterious book to commoner Liu Bang's strategist Zhang Liang (张良), who helped Liu Bang become the one king to rule them all. This book resurfaced almost sixteen hundred years later in the hands of Liu Ji (刘基), the strategist of Zhu Yuanzhang (朱元璋), the wandering mendicant monk who became the founder of the Ming dynasty.

"At his deathbed, Liu Ji forbade his sons to study the mysterious book and told them to give it to the emperor," said David. "No one knew where the book went after that. This book originated from during Shang times and helped establish the Zhou dynasty. The Sage Yellow Stone who helped establish the Han dynasty is a secret clansman in my lineage book. My father Huang Shi (Yellow Stone) is named after him."

A thought struck Victoria like thunder on a sunny day. She realized she might have solved a big piece of the puzzle in David's epic story. Victoria had an idea of the whereabouts of this mysterious dynasty-building book. She felt emboldened by the power of secret knowledge. A surge of confidence gave her the courage to be more assertive.

"Which brings us back to the question of what is the stopping-of-the-sun?" asked Victoria, steering David's random walk down the garden path back to her original objective. "Why do you, the scion of a king-maker lineage, risk your life to protect me? Who are the people that are trying to kill me? And who am I to enjoy this kind of unwanted attention?"

"You're good at solving puzzles," said David. "Why don't you take a stab at it?"

"According to you, I am a princess, so my parents must be royalty," said Victoria. "I have a *huang* jade pendant which is the ancient symbol of royalty, and all those Oracle Bone glyphs on the pendant tell me that I am connected to that era. You told me about secret clans of unbroken lineage starting all the way from the Shang era. They're powerful; they build and destroy dynasties; and they have to protect me because I am going to save humanity when the sun stops in the sky. I get it—I'm Supergirl. Krypton blew up. My royal parents arrived on earth during Shang times. They put me in a time machine and sent me to the future to save the world, but Lex Luther wants to stop me, so here I am."

"That's actually not a bad guess," said David, smiling at Victoria's answer. "Do you ever wonder why you have never been sick or how you survived that horrific car crash with nary a scratch?"

Victoria shrugged.

"Come on, no one is that lucky," exclaimed David. "The truth is that, you have self-healing powers. You also heal people around you. I was having a pretty bad headache until you took it from me."

"Am I a mutant," Victoria asked, "like in *X-men* or *Heroes*?" She was not sure if David was pulling her leg.

"I don't know," said David. "The answer awaits us in China. Although I have always suspected that you have healing powers, I wasn't sure until Michael and Angela confirmed it. You know that your adoptive parents were from the Ukraine but you don't know that they were victims of the Chernobyl disaster, when a nuclear accident contaminated a large area with radioactivity. Their families all worked at the site. Michael and Angela lost everyone to the catastrophe. It was a bitter experience. They also developed cancer but were lucky to have survived after treatment.

"I thought Michael and Angela had a clean bill of health when they agreed to take you. It turned out that they both suffered relapse and metastasis. They had the most virulent form of Non-Hodgkin Lymphoma. Their oncologist told them to just enjoy their time together and prepare for the inevitable. After having you for several months, they went back for a checkup. Their cancer was miraculously in remission. After one year, they were both a hundred percent cancer free. You were easily tired and slept a lot that year, but returned to normal afterwards. How do you explain that?"

"No wonder in my dad's letter, he said without me they would have died long ago," said Victoria, her eyes tearing up by being reminded of the death of her parents, the only ones she had ever known. Now she understood why her parents didn't want to talk about their past.

It was dark by the time they arrived at Truffle Pigs. As they walked into the restaurant, Victoria could smell the sweet aroma of roast pork wafting in the air. She had to remind herself the etiquette lessons she learned from her mom, to always nibble and never wolf when dining in public.

After sitting down at their reserved table, Victoria began to have an eerie feeling of unease. She soon found out why. In the shadow of a dark corner, a glass eye was staring at her. It was the creepiest feeling to be stalked by a homeless Toronto man with a glass eye in the middle of the Rocky Mountains. Victoria tugged at David's shirt to alert him. Then she noticed something wrong with David. His face had turned white as a sheet. Two men in black overcoats had just walked in. One of them was carrying a briefcase.

"Stay calm; don't make any sudden moves," whispered David. "Walk slowly out of the backdoor. Moira will be there with the car, just get in and go."

"But what about you?"

"I'll be fine," said David. "Moira knows how to get me. Go now."

Victoria was glad to see Moira and the F011 waiting outside with the car door opened. The moment she sat down, the car sped towards David who had just bolted out of the front door in a hundred metre dash. The two men in black were chasing after him. Victoria could see a shadow emerge from the backdoor. The Toronto one-eyed man was chasing after her too. How did these people know they were coming to Truffle Pigs? How did the homeless guy get there even before they did? Who were these people?

"Moira, drop a guided Mark-6 EM mortar on the parked cars," ordered David as he hopped into the future vehicle.

A loud humming noise issued from the F011. The lights at the parking area promptly went dark. The front panel of the parked cars started spewing sparks and smoke. The lights in Truffle Pigs dimmed and flashed trying desperately to stay alive. The men in black were running to their SUV but stopped in their tracks. Sparks were shooting out of the briefcase as well.

"Damn," Vincent cussed. "We need spare parts."

The one-eyed man stood in the shadows watching David and Victoria escape. Sparks and smoke were coming out of his glass eye, but as if by magic, the glass eye repaired itself.

The F011 sped through the Rockies in the dark. They did not stay at Pan Pacific Hotel. Booking ahead and being predictable was not a good idea. They walked into a hotel and paid cash for their rooms. The passport and Chinese ID card for Victor So arrived by special delivery the next day, and they boarded a container ship leaving port before New Year's Eve. They were finally on their way to China.

CHAPTER 31

Story of the Stone

"If you have tears, prepare to shed them now!" Huang Shi (Yellow Stone) heard the saddest news as he stepped off the long distance bus at the Yellow County Bus Depot. It was a long and bumpy ride from Anyang. The rickety bus would continue its jaunty journey to Qingdao (青岛; green island), the old German colony by the oceanic coast of the Shandong peninsula.

Huang Shi came home to a piteous sight. Everyone around him was weeping uncontrollably. The news broke that the people's beloved Premier Zhou Enlai (周恩来) had passed away. It was the 8th day of January. The year 1976 had begun badly. Gone was the anchor of the people in the national turmoil known as the Cultural Revolution. What was to become of China? What hope was there for Huang Shi?

The Cultural Revolution was half way past its tenth year. While the extreme turbulence of the early days had calmed down, the opportunists who had unscrupulously damaged the ruling Chinese Community Party, the nation, and its people were still in power. Who knew what they would do next?

Everyone had such high hopes in the beginning of this popular movement. The people would enjoy unprecedented freedoms. The masses would be in charge of making their own decisions and running their own lives. The young and restless had free rein to rebel against traditional social mores and conservative institutions long resistant to change. It was said that the old guards of the revolution had become corrupted by their privileges and their bureaucratic tendencies. They were dragging their feet in reform and had become enemies of the people. Established order must be overturned by a new revolution—the Cultural Revolution. The fossilized conservative Chinese culture and the authoritarianism of the Confucian social hierarchy must be abolished forthwith.

All levels of government, schools, news agencies, art institutes, cultural organizations, and social services fell one by one to the chaos of the masses. Anarchy reigned supreme. The excitement of revolution fanned by ignorant fanaticism and youthful exuberance unfettered by rules or reason left the country in ruins.

The first victims of the Cultural Revolution were many of the early revolutionaries who had risked life and limb to usher in the People's Republic of China. For example, General Peng Dehuai (彭德怀), one of Chairman Mao's comrades since the early days of the revolution, the Commander-in-Chief of the Chinese forces in the Korean War, and later the nation's Defence Minister, suffered repeated public humiliation and torture at the hands of the Red Guards. He died ignominiously in prison.

Liu Shaoqi (刘少奇), the President of China, the number three person after Chairman Mao and Premier Zhou, likewise was attacked by the Red Guards and died while in prison. His famous wife, the nation's First Lady, the beautiful, cultured, and elegant Wang Guangmei (王光美), who spoke multiple languages and had a Master's degree in physics, was imprisoned for twelve years, most of it in solitary confinement.

Deng Xiaoping had the special honour to have been purged twice during the Cultural Revolution. He was sixty-five years old when he was banished from the central government to work as a common labourer in a factory. His son Deng Pufang became a paraplegic after falling from the fourth floor of a building while trying to escape from the Red Guards. Since no hospital dared to treat Deng Pufang, Deng Xiaoping had to take care of his son with no professional help. Deng's own younger brother committed suicide rather than endure humiliation at the hands of the Red Guards.

Deng was at retirement age; yet he did not give up and die a bitter old man. In the face of impossible odds, Deng steadfastly refused to concede defeat, rising instead from the ashes at the ripe old age of seventy-two to lead the nation to greatness, helping even those who had humiliated and tortured him and his family. Deng maybe short in physical stature, but he is a giant in Chinese history.

If the highest echelons suffered such indignant fates, it followed that hardly anyone else was spared. It was only a difference in the degree of severity. The lucky ones got off with humiliation and the loss of position in the government, others were locked up in solitary confinement and left to rot. The less fortunate ones would lose their lives after repeated public humiliation and torture. Intellectuals, scientists, artists, scholars, and students were sent to the countryside to work in the farms and collectives.

By the end of the catastrophe, national census showed that almost a quarter of the nation's population was illiterate or semi-literate. The nation ground to a standstill for ten years and was left in the dust while the rest of the world forged ahead.

The official history of the Chinese Communist Party acknowledged that the Cultural Revolution was a grave mistake. It was a self-inflicted wound. It became a disaster for the Party, the nation, and its people. It was through introspection and acknowledgement of the mistake that the Chinese Communist Party could examine what happened and declare "Never Again!"

The government's own statistics showed that during the Cultural Revolution, 2.3 million government officials were investigated. An even larger number of government officials who were not officially investigated were subjected to wrongful prosecution. Seventy-five percent of central government officials, vice-ministers, provincial/regional ministers and their officers were investigated. More than sixty thousand government officials died because of these wrongful prosecutions. There were almost twenty thousand cases of wrongful conviction. Many important government institutes suffered paralysis during the ten years of the Cultural Revolution. It was under this perennial dark cloud that Huang Shi had survived the past decade.

This is the story of David Huang's father Huang Shi, who was born Huang Taishan (黄泰山; Tai Shan, located in Shandong, is one of China's most important mountains) in Yellow County of Shandong province about nine years before the establishment of the People's Republic of China.

Shandong is a blessed province. Besides being the home of Confucius and Confucianism, it has fertile soil, abundant seafood, deep-water seaports, and it is located strategically between Beijing and Shanghai, the two largest and most important cities of the country. Yellow County was also a short boat ride to Manchuria. Being thus endowed with geographic advantages, it made sense for the people of Yellow County to engage in the trading business.

Huang Taishan's family had been merchant-traders for many generations. His family branch had broken off from the main clan to do business in Manchuria. But then the Japanese took control of Manchuria on September 18, 1931, the Day of Infamy for China. All Chinese people remember 918, just as all Americans and Chileans remember 911, all Iranians remember 819, and all Congolese remember 117.

Refusing to cooperate with the Japanese occupation, Huang Taishan's grandfather moved his family back to Yellow County. Hard times fell upon the family, the grandparents died, and Huang Taishan's parents became the last surviving members of the branch. Huang Taishan was born in the year 1940. China had been at war against full-blown Japanese invasion since July 7, 1937. Europe was in flames, and America would soon join to make this a worldwide conflict.

Life was still bearable until around the time when Huang Taishan was three years old. His father went to Manchuria with a large consignment of fabrics from Hangzhou (杭州) but never came home. By that time, the Japanese had occupied all of the coastal areas of China, including Yellow County. In order to survive, Huang Taishan's father could no longer choose where or with whom he would do business, and he had buyers in Manchuria.

The war actually wasn't going well for Japan. The Americans had sunk four Japanese aircraft carriers in the Battle of Midway. Japan would never recover from the loss. The expected quick and easy subjugation of China turned into a debilitating stalemate. It was death by a thousand cuts. In occupied areas, Japanese forces did their best to keep the peace, but occupiers shall not have peace. Nationalist spies and Communist guerillas constantly harassed the Japanese occupiers. Hence from time to time, the Japanese forces felt obliged to conduct reprisal operations, which when not exercised with restrain, often degenerated into orgies of fire and fury, the opposite of peace.

The Japanese had hoped that Generalissimo Chiang Kai-shek, the leader of the Nationalist government of China, would negotiate for cessation of hostilities. Unfortunately, the Nationalists under Chiang Kai-shek and the Communists under Mao Zedong made nice after the Xi'an Incident, during which Chiang was arrested by his own generals, ironically gaining release after the Communist's mediation. The Nationalist political party Kuomintang (KMT) and the Chinese Communist Party (CCP) had cooperated during the 1920s in their war against the warlords. This time, they cooperated again in the war against the Japanese.

No one knew exactly what happened to Huang Taishan's father; the consensus was that he got in trouble with the Japanese. That sort of thing happened during war, especially when the Japanese was implementing the *Three All* policy of 'burn all—kill all—loot all.'

In life and particularly in war, one does not own anything one cannot defend. Huang Taishan's father could not defend his freedom to trade, could not defend the ownership of his goods, and could not defend his right to live. He owned nothing, he lost everything, and his family lost him.

CHAPTER 32

The One-eyed Slave

With the disappearance of his father, the loss of the goods, and the materialization of loan collectors at the door, Huang Taishan's family was immediately plunged into the pit of perdition. At a time when no one had the means to help, Huang Taishan's mother committed suicide by jumping into a well.

Fast forward past the days when Huang Taishan had to withstand the shock of losing both parents, crying to sleep, and begging for food to stave off starvation, his luck finally turned around. A young slave-girl of a rich man's daughter saw the desperate boy when she passed by on an errand. She decided to help. Everyone in the neighbourhood knew the kind-hearted girl as Sister Hao (好姐: hao-jie), or Good Sister. That was the only name she had, and she would never acquire a name by marriage. The slave-trader had blinded her in the left eye to condemn her to a lifetime of slavery. She was however, more human than any of the book-toting saint-quoting selfish hypocritical egotistic erudite person Huang Taishan had ever known in his life.

Without expecting even a word of gratitude, Sister Hao fed Taishan, cleaned him up, and took him to a Huang family, which took him to another until eventually he was adopted by a distant relative—an uncle of the Huang clan—who happened to have three daughters but no son. Huang Taishan was lucky to have survived, because during wartimes, no one had food to feed extra mouths. When Huang Taishan joined his new family, they gave him a new name, Huang Shi, or Yellow Stone. Huang Shi wanted to find Sister Hao when he was older, but learned that she had moved with her mistress to Shanghai and then to Hong Kong. No one could have known that the fate of the universe hanged in the balance when a one-eyed slave-girl called Hao saved the day.

Huang Shi's uncle and adoptive father happened to be the scion of the unbroken Huang lineage and keeper of the secret. He had been pondering the problem of passing the secret to one of his daughters. Huang Shi solved that problem to his adoptive father's immense relief.

Huang Shi's adoptive father had given up trading some years ago, and had used his savings to buy himself a small plot of farmland, which kept his family surviving sufficiently if not sumptuously. His adoptive father was naturally sympathetic to the anti-occupation guerillas, many of whom also happened to be communists.

During the Chinese Civil War, the decisive Huai Hai (淮海) Campaign was fought around the city Xuzhou (徐州) of Jiangsu (江苏) province to the southwest of Shandong. In about two months at the end of 1948, the crème de la crème of Chiang Kai-shek's Nationalist army was annihilated. This was despite the fact that the Americans had armed and trained them. The Nationalist army also enjoyed numerical superiority, had several fully mechanized army corps with tank divisions, supported by an air force and a navy, and were led by the best and most experienced generals, who had graduated from the elite Whampoa (黄埔) Military Academy, the Chinese equivalent of West Point.

The Communist side did not have any of the above, and the effective field commander in the main battle, Su Yu (粟裕), the son of a peasant, did not attend any military academy. He was such a nobody that none of his opponents knew they were fighting a major battle against him. Even after the establishment of the PRC, historians had a hard time learning who was responsible for this brilliant campaign until decades later, when they were able to study original documents from the archives. Su Yu was humble to a fault and refused to take credit.

There were rumours that Su Yu received help from a secret Confucian clansman. Clues were hidden in plain sight, but the rumours remained unproven. In truth, the Communist army was supposed to be a band of guerillas unschooled in conventional warfare and incapable of conducting one, especially a major battle that involved eight hundred thousand soldiers on the Nationalist side and over six hundred thousand soldiers on the Communist side. Su Yu somehow transformed peasants in straw sandals into a feared and lethal fighting force.

Furthermore, the Nationalist generals knew that it was a battle they could ill afford to lose. They were the last line of defence for their capital Nanjing (南京; south capital), also known as Nanking. A cornered beast was a dangerous beast.

Nevertheless, the Communist army cut off, encircled, starved, and vanquished the entire Nationalist army in the theatre. The de facto commander of the campaign on the Nationalist side was the famous general Du Yu'ming (杜聿明), who was a first class graduate of Whampoa, and who had fought in the Burma Campaign under General Joseph Stilwell. He would later become the father-in-law of winner of Nobel Prize in physics Professor Yang Zhen'ning (杨振宁) of Princeton's Institute for Advanced Study, whose claim to fame was the highly successful Yang-Mills Theory. Communist soldiers captured Du when he tried to break out. Despite all the paper advantages, the Nationalist army lost the will to fight when there was no food.

Days after the conclusion of the Huai Hai Campaign in early January 1949, the Nationalist army defending the isolated Beijing surrendered. Manchuria had fallen before the commencement of the Huai Hai Campaign. Chiang Kai-shek had no significant armies to stem the Communist tide. Chiang's vertiginous fall from grace took only months. America backed the wrong side in this civil conflict.

On the 1st of October in the same year, the People's Republic of China (PRC) was born. Politicians in America started blaming each other for having lost China. Most Chinese didn't know that America owned them. Soon Cold War mentality and McCarthyism took hold of the American psyche. With no possibility of a quick rapprochement accompanied by reciprocal respect, China went her own way, as if the Americans had never come.

There were many theories for the astounding result of the Huai Hai Campaign. Some said that it was Chiang Kai-shek's poor support for his generals. Others attributed the victory to the fine work of the Communist General Su Yu, who rather than followed orders to take diversionary action south of the Yangzi River, obstinately stayed in the central plains to fight a decisive battle against the ostensibly much stronger and advantageously deployed Nationalist army. Others mentioned how the Communist generals worked together as a team towards a common goal, while the Nationalist generals had private agendas.

Almost everyone agreed that an important factor was that the farmers from around the countryside, even all the way from Shandong took part in supplying the Communist army. The Nationalist side on the other hand depended on airlifts after being cut off and encircled. When poor winter weather did not permit flying, the Nationalist soldiers starved.

Huang Shi's adoptive father was one of the farmers who supplied the Communist army. He backed the winning side. Coming from such a family, Huang Shi was looked upon with favour. Life was going to be fine from then on. Huang Shi didn't disappoint either. He was a top student and he worked hard to get into the best school there was.

Huang Shi inherited the secret lineage book when he was nineteen years of age, before he left home to study at the prestigious Peking University. At the time, Huang Shi thought that the lineage and oracle was just fuddy duddy mumble jumble of ancient peasant superstition, but he humoured his father, who treated the matter with utter seriousness.

Between 1959 and 1961, when the country was hit by poor harvests and Shandong was one of the harder hit places, everyone in the clan ate a bit less to make sure Huang Shi had enough nutrition. Huang Shi was the pride of the family. When Canada's future prime minister Pierre Trudeau, also known in history as Trudeau the Elder, the Rock of Canada, visited China in 1960, he came to Peking University and Huang Shi was one of the star students selected to meet him.

At the time, Canada had excess grains and its government was eager to supply it to China, on loan if necessary, to alleviate its food shortages. But that did not happen because of a Western embargo on China. Decades later in 1996, liberal politician Madeleine Albright, at that time the U.S. ambassador to the UN, and future Secretary of State for the Clinton administration, clarified the position of Western embargoes. When Leslie Stahl of *60 Minutes* asked Madam Albright for her opinion about a UN report, which said that five hundred thousand Iraqi children had died as a result of five years of U.S. embargo, she famously replied, "We think the price is worth it."

China had no illusions that it was on its own. Sometimes the harshest attacks came from the left. Your 'friends' stab you in the front. Rationing and the sharing of food helped the country pull through those two difficult years. No doubt, there were suffering, displacement, malnutrition, ill health, and unnecessary loss of lives.

It could have been easier if China's relationship with the USSR had not soured while the West and their allies had the country under embargo, but the people of China did not want to surrender their dignity and independence, which they had earned with heavy payments of blood and suffering.

China had survived worse famines in the past, many with devastating consequences. Its resilient people survived this one by tightening their belt and biting the bullet. They thus dashed the hopes of those who had prayed every day for the country to fall apart and to become easy prey again.

China had completely recovered by the time Huang Shi received his doctorate from Peking University. Huang Shi had by then become a top expert in Shang culture and its ancient writing, the Oracle Bone script.

Huang Shi was riding on a wave of good fortune in 1966. At the beginning of the year, Huang Shi received approval for his Communist Party membership application and he finally became a full member.

Not long after, Huang Shi also received approval from his Party superiors to marry his university sweetheart Lin Yu (林玉), a graduate student in civil engineering who was a war orphan from Tang Shan (唐山; Tang Mountain), a city about 160 kilometres (100 miles) to the east of Beijing by the coast of Bohai (渤海; Bo Sea). They celebrated the news of Lin Yu's pregnancy in early May. On that same day, Huang Shi learned of his posting to the Anyang branch of the Institute of Archaeology, one of many institutes of the Chinese Academy of Social Science, as a senior researcher under the renowned and respected Professor Zheng.

Professor Zheng was the first generation of archaeologists to have graduated from Peking University after the establishment of the People's Republic of China. She would eventually spend her entire career in Anyang. At the northern edge of Anyang is Yinxu, the archaeological ruins of the ancient Shang capital Yin in the Province of He'nan. That is the only place in the world where a Shang and Oracle Bone scholar would want to be. For Huang Shi, life could not be more perfect.

Huang Shi planned to take Lin Yu back home to give birth, where she and the new baby would be well taken care of, while he headed out on his own to organize his long term working and living arrangements in Anyang.

On the 25th of May, Huang Shi and Lin Yu said goodbye to their friends, colleagues, and professors, and walked out of their dormitory in the campus of Peking University with a cartload of books and suitcases. Before the bus arrived, Huang Shi heard a commotion. It came from the direction of the refectory. People were running in that direction as well. Huang Shi couldn't help being nosy, and ran over despite his pregnant wife's protests.

On the wall of the refectory, Huang Shi saw a Big Character Poster (大字报; da-zi-bao), a means by which anyone could publicly express their opinions on anything—theoretically—sort of like a Twitter account nowadays on the internet. It was presumably how the public exercised the freedom of speech. This particular poster brazenly attacked the leadership of Peking University.

One might think that Peking University would be spared. After all, Chairman Mao was an assistant librarian at the university before he became one of the founders of the Chinese Communist Party. Was nothing sacred? Unfortunately, the Cultural Revolution had arrived and schools would be one of its first battlegrounds. The dark clouds of chaos would soon enshroud the nation. None could escape unscathed. From that day, Huang Shi's life began to unravel.

When he was studying at Peking University, Huang Shi had the opportunity to befriend some foreigners, many of whom were idealists who had chosen to stay in China to help the development of the Chinese society under communism. While there were thousands of them, Huang Shi became good friends with a few; they included George Hatem, Sidney Shapiro, Israel Epstein, Sidney Rittenberg, and Joan Hinton.

They were all distinguished people in their fields. Hatem was a medical expert, renowned for his work on leprosy and venereal diseases. Shapiro translated the *Water Margin,* one of China's four classics. Epstein was the editor of the magazine *China Today*. Rittenberg was a senior government official in media, journalism, and broadcasting. Hinton, one of only a few female scientists to have worked on the *Manhattan Project*, bred awesome cows with her husband.

Huang Shi thought of them as teachers and mentors. He made friends with them to learn about the world outside of China and to improve his English.

Huang Shi noticed that most of his foreigner friends were at first excited and supportive of the Cultural Revolution. After all, what could go wrong with more freedom for individuals and less government bureaucracy?

As the revolution spun out of control, the freedom to be diligent and responsible devolved into the freedom to be ignorant and cruel. The freedom to be polite and respectful became the freedom to be rude and ruthless. Moreover, one must always defend to the death the freedom to tell lies, the freedom to impose lies on others, and the freedom to deprive others of freedom.

Even Huang Shi's foreigner friends were not spared. The people in power locked up Rittenberg and Epstein, denounced Hatem, and placed Shapiro's wife under house arrest. Hinton and her family had to leave their beloved farm. Huang Shi had dark forebodings that winter was coming. He didn't figure on the frosty frozen winter to last for ten frigid years.

CHAPTER 33

Gathering Clouds

In the second half of the year 1966, Huang Shi's luck took a turn for the worse. After arriving home and holding the traditional wedding dinner, Huang Shi's wife Lin Yu had a serious case of food poisoning, immediately followed by a most unfortunate miscarriage. The fetus was already three and a half months old. It was a girl. Lin Yu was inconsolable for months until she learned that she was pregnant again.

Huang Shi's posting at Anyang became a bureaucratic nightmare. The Institute of Archaeology was under attack. Professor Zheng was too frazzled trying to fend off the Red Guards, and there were no officials to deal with the paper work required for Huang Shi's job posting. The Red Guards didn't care; the people's democratic authority was supreme.

Finally, Professor Zheng sent Huang Shi a telegram, apologetically telling him that no one was in charge for confirming his employment, cryptically bidding him to bide his time. Huang Shi was devastated because by then, he couldn't go back to Peking University, and it was not safe to be too much of a scholar anywhere. The best idea for Huang Shi was to lay low in his hometown village at the edge of Yellow County as the son of peasant parents.

Huang Shi's adoptive father exhausted all the good will and connections he had accumulated through the years to get Huang Shi a job at a nearby factory, where they produced the famous

Long Kou vermicelli. He could make a living and be safe as long as he kept his mouth shut.

Huang Shi told himself this was temporary. He could always research and write in his spare time. Many Party members, government officials, and academics had suffered much worse. At least he was with his family and not being paraded in the streets or locked up in a cell, or suffering any of a number of other insufferable indignities. Huang Shi had faith that all these would blow over eventually and things would return to normal. Things always return to normal. China had gone through much worse in the past and survived. So he clammed up, put his head down, and went to work as a common labourer in a vermicelli factory.

After learning all he needed to learn about the job in approximately half a day, Huang Shi settled down into the routine. Unlike the satisfaction of deciphering Oracle Bone scripts or the excitement of making archaeological discoveries, factory work could be repetitive and monotonous. Huang Shi convinced himself that at least his life was calm, and a child was on the way, so he could look forward to that.

One day, while he was taking a regular tea break at the factory, a neighbourhood boy almost flew past the front gates on an oversized bike he obviously borrowed in a hurry. The news on his face was worrisome. Lin Yu tripped and fell while walking home from the market. She had gone into labour. She was two weeks early and was bleeding badly. The village midwife came and was doing her best. Huang Shi should take the rest of the day off and rush home.

Huang Shi raced through the narrow alleys and the shortcuts on his bicycle, ringing his bell all the way. A brisk bike ride from the factory to home would take fifteen minutes, if Lin Yu was in danger, wasting half an hour for him could mean the difference between life and death.

When Huang Shi arrived at the old shed he called home, it was eerily silent. He rushed into his bedroom without noticing who was in the house. He knew there were people standing in the dark corners with downcast eyes and pale faces. No one made a sound. There were water basins, overturned stools, and bloody towels strewn about on the floor. Red sheets covered the still body of Lin Yu.

Red was the festive colour of China. Even the Cultural Revolution couldn't take that away. After all, China's flag was red. Huang Shi didn't remember having red sheets at home. It was Lin Yu's blood. It was the sacrifice to the gods so that his son would live—his son the newborn babe who was wrapped in swaddling cloth and sleeping peacefully beside his dying mother, dreaming of a future in which he would be a valiant knight in shining armour, rescuing a beautiful princess from the jaws of the Leviathan.

"I'm so sorry," sobbed the midwife. "I've tried everything, but …"

Huang Shi ignored her and knelt beside his wife. He didn't want to waste time listening to the midwife's blubbering babble. Every second he could spend with his dying wife was more precious than all the tea in China. If he traded everything he owned, he could not buy even a single second for Lin Yu. Huang Shi felt the helplessness of humanity in the face of mortality.

Lin Yu had held on so she could see Huang Shi for the last time. She would not leave without saying goodbye. Lin Yu had not a drop of blood left on her face. Her tears had dried at the corner of her eyes. She was poignantly beautiful even as she was dying.

"You're back," whispered Lin Yu. "Look at our lovely baby."

"You're going to be fine," said Huang Shi, but the tears ignored his poor attempt at prevarication and cascaded from his lying eyes.

"We have a daughter in heaven, I'm going to be with her soon," said Lin Yu. She never forgot the child that she lost from her miscarriage. "Let us call our son Didi (弟弟)—little brother, so he knows he has an older sister. We'll be waiting for you and Didi. Live a good life for me and look after yourself—I'm so tired—I have to rest now."

Lin Yu smiled and closed her eyes to dream the dream that would end the heartaches and the thousand natural shocks of life.

Thereafter, Huang Shi's parents and sisters helped raise his son, while he worked at the same factory going through the motions of the same routine day in and day out, still waiting for that call from Professor Zheng, which never came. Days turned into weeks and weeks turned into years, Huang Shi sipped his teas and swallowed his tears, counting the tick-tocks of the factory clock that made up his dull days, and then one day he found, ten years had passed him by—ten years wasted and ten years closer to death.

On the Winter Festival of the year 1975, during which the Chinese people gathered for the most important traditional family dinner of the year, a neighbourhood kid came running from the local Post Office shouting all the way that Huang Shi had a long distance call from Professor Zhang. The call would take him immediately to Anyang, but not for the reason he had hoped. He went there to learn of the bitter future that awaited humanity.

When Huang Shi returned home on the eighth day of January 1976, he was burdened with the knowledge of how the world would end. He dared not divulge the terrible secret to another soul, for fear it would violate some inviolable law of nature. The survival of humanity was at stake. Huang Shi must be prepared to sacrifice everything.

Chapter 34

Year of the Dragon

During the Pacific crossing, Victoria critically analyzed all that she had learned on this trip. She even verified her healing powers. She learned that David Huang had a rather tragic childhood because he never knew his mother. His father Huang Shi was a brooder when he was not working at the vermicelli factory. Everything changed for David on the day of the Winter Festival of 1975, when he was eight years old.

In China, people celebrate the Winter Festival on the day of the winter solstice, known traditionally as the 'Extreme of Winter' (冬至: dong-zhi). It is one of the most important festivals of the year, when families get together for dinner, similar to Thanksgivings in North America.

David remembered that his father hurriedly left that evening, and came home on the 8th of January, 1976. Premier Zhou Enlai passed away on that day. David had never seen so many people spontaneously weep at the same time. It was as if China was hit by the holy plague and God had struck down the first born of every home. When Huang Shi entered his dimly lit shed, everyone waited for him to deliver the good news. But Huang Shi said nothing. He just walked inside his bedroom and locked himself in.

This was the Year of the Dragon. It should have been an auspicious year for China, the land of the Dragon. No one had expected the three stars to fall from the sky that year. First, it was Premier Zhou Enlai. Then Grand Marshal Zhu De (朱德), the founder and Commander-in-Chief of the People's Liberation Army (PLA), passed away on the 4th of July. Finally, it was Chairman Mao's turn to make the grand exit on the 9th of September.

In April, the *Gang of Four* purged the rehabilitated Deng Xiaoping again. In May, disaster struck David's family when his father vanished from the face of the earth. A month later, his grandfather died from a broken heart. In July, twentieth century's most destructive earthquake flattened Tang Shan, the hometown of David's mom. Six hundred and fifty thousand people went to bed and never woke. It was carnage of apocalyptic scale. Sic transit gloria mundi—thus passes the glory of the world; dum vivimus vivamus—while we live, let us live.

Huang Shi was spared the horrors of Tang Shan, but what he had witnessed was infinitely more horrifying. Huang Shi finally decided to bring his father and son into his room and told them what had transpired. It was all a bit too much for the eight years old Didi, but Huang Shi had no choice.

"What happened at Anyang?" asked Huang Shi's father. "Has the Oracle arrived?"

"I'm afraid so," said Huang Shi. "At first, I thought Professor Zheng urgently needed my help to deal with a discovery. When I got there, I found that a young man, maybe just a teenager, had insisted on meeting me in return for showing Professor Zheng a discovery of the ages. The young man was very convincing; he knew the secret clan password, and he knew that I am a secret clansman.

"He said that his wife had just given birth to a baby girl at sunrise on the day of the Winter Festival. The young man explained who they were and showed me the Oracle carved on jade pieces. He insisted that I must hide their daughter and keep her safe. She must grow up in safety because the world will depend on her."

"Who are they?" asked Huang Shi's adoptive father. "How do you know they were telling the truth?"

"They're not of this world or maybe not from our time. The young man demonstrated his god-like abilities," said Huang Shi. "More importantly, he has a briefcase with equipment inside that has astonishing powers. He showed me the future with it; a future he said would happen if his daughter fell into the wrong hands. When he opened the briefcase, a light shone into my eyes. I felt that I went to another world. Everything was very real. I saw how the world would end. It's horrifying and beyond description."

"Where is the baby?" asked Huang Shi's father.

"The baby is still with her parents. They say she is too small and not ready," said Huang Shi. "They will arrange for me to pick up the baby in May. The good news is—I am getting my research job after all. I will return to Anyang in April when excavation season starts.

"The young couple gave me an advice," Huang Shi added. "Whenever you see two men dressed in black, and one of them carries a briefcase, run. They are enforcers known as Watchers or Maleks. They are messengers and henchmen of a powerful alien entity known as Dias. They must not get the baby."

When April came along, Huang Shi bid farewell to his family and left for his new job at Anyang. Huang Shi didn't know that he had said his last goodbye to his adoptive father, and he would not be a part of his son's growing up.

CHAPTER 35

Rare Treasures

David remembered this day vividly. In the evening of May 17, 1976, the family got a call from Huang Shi. He reported that everything was going well in Anyang, and that the strange young couple did show up again with their baby girl. The young man completed his end of the bargain by providing specific information of where rare treasures were buried. Professor Zheng and her staff drilling at an unexpected location down to an unprecedented depth yielded evidence of an undisturbed royal tomb. Everyone who was at first skeptical was now astonished and excited. Huang Shi said that he expected the young couple to entrust him with the baby in a day or two. He would immediately return home to work out how to hide the girl.

In the following days, the newspapers would declare a major discovery at Yinxu. The Huang family excitedly read about this royal tomb coming to light and the wealth of treasures being unearthed. There was a large cache of martial implements, leading the scholars to think that the tomb belonged to a prince, but further discovery of a cache of female artifacts indicated that the tomb belonged to a princess or a queen. When workers lifted bronze axes and large bronze vessels from the ground, Professor Zheng was able to confirm the name of the tomb's owner. It was the famous queen Fu Hao of King Wu Ding. Her name had earlier shown up in over two hundred Oracle Bone records. We know her as Queen Zia.

No tomb raider ever robbed Queen Zia's tomb for a special reason. The distraught King Wotan got a suggestion from his trusted one-eyed official Huang Yin to build a temple for his beloved Queen Zia right at his palace, and to bury her underneath the temple. That way, Queen Zia would be close to the king, and her spirit would both protect and comfort him. Since the Shang built their houses on a foundation of rammed earth, and they never built royal tombs within their palace compounds, the tomb raiders and archaeologists would stop when they reached rammed earth, thinking they would find nothing underneath. This protected Queen Zia's tomb until a mysterious young man gave specific drilling instructions to Professor Zheng.

David's family had expected Huang Shi to come home but he never showed up. They got no answers from their calls to Anyang. Huang Shi had vanished without a trace. Huang Shi's father went to Anyang and searched high and low, but no one had seen his son. He searched hither and thither. He sought the mysterious young couple and their baby; they were nowhere to be found either. The poor old peasant from Shandong spent a month in the neighbourhood with nothing to show for. In despair, he bought a bottle of hard liquor known as 112 proof *Erguotou* (二锅头; double distilled) *baijiu* (白酒; white liquor), sat underneath a willow tree, and drank his sorrows away.

After finishing the bottle, the old man fell asleep and went to a wonderland of dreams, where he saw his adopted son for the first time, and he wept with joy at his good fortunes to have received such an excellent gift from heaven. The next morning, passing peasants found the body of an old man slumped against the willow tree. The search for Huang Shi had ended with the death of his father. There is an old saying in China: "Good luck never comes in pairs while misfortune never strikes alone." It's Murphy's Law with Chinese characteristics. Didi was a young lad but perforce became the man of the house, inheriting the secret lineage book at the same time.

Didi followed his father's footsteps to become a scholar in the study of antiquities. Like his father before him, he was a top student, graduating magna cum laude from Peking University, and was accepted for post-graduate studies at the University of Toronto's Department of East Asian Studies. The whole clan dug up all the money they had to send David abroad. Their hopes now rested on Didi's shoulders. They did not however expect Didi to become David the shady art-and-antique dealer.

When Chinese people started to sacrifice their youth, their health, and their environment for American paper notes, without which they could not buy oil and technology to fuel their economy, David went back to China to seek opportunities, and he found gold hidden in plain sight. Items that David culled at Beijing flea markets could easily fetch a hundred times their costs in a trendy shop in New York City. But even that was small potatoes when David got lucky.

Like most antique dealers, David revelled in regaling others with tales of extraordinary discoveries. He told Victoria how he once bought a rare vase at Beijing's antique shopping area Liuli Chang (琉璃厂; Glaze Factory). It was a product of the legendary Southern Song Xiunei Si (修内司; Department of Internal Manufacture) Royal Kiln in Hangzhou.

What a day David had at Sotheby's!

"Lot 116, ladies and gentlemen—a flawless vase of Southern Song Xiunei Si Royal Kiln in the shape of a Shang-Zhou vessel, coated with crystalline crackled glaze over stunning jadeite green. This item, discovered in the storage of an old antique shop which once belonged to the Qing Prince-Regent Chun (醇亲王), Zaifeng (载沣), father of Puyi (溥仪) the last emperor of China, is extremely rare in its unprecedented perfect condition. May I start at five hundred thousand?"

David sold the vase for eight million U.S. dollars after a heated bidding war. He had bought it for RMB 850 yuan, about a hundred U.S. dollars in those days. A mysterious bidder in black Armani suit wearing a custom designed diamond-studded eyepatch with bidding paddle number eleven won the bid.

David also recounted how he found an extremely rare mint condition royal porcelain cricket-can from Ming Dynasty's Emperor Xuande (宣德) at Beijing's famous antique market Panjia Yuan (潘家园; Pan Family Garden).

David sold the can at a Christie's auction for three million U.S. dollars. David had a notebook full of deals like that. He made so much money he almost felt sorry for the people who didn't know they had gold mines in their possession. That was raw free market capitalism for David. He made the money fair and square. Peter Minuit didn't feel sorry for the natives who sold the Manhattan Island to him for sixty guilders. Both sides were happy with the deal. Sixty guilders can't buy one square-foot (a tenth of a square metre) today while the vendors didn't actually own the real estate.

Thirteen years ago, while David was celebrating Winter Festival at his renovated Long Kou (Dragon Mouth) home with his grandmother and his cousins' families, a most unexpected visitor came for dinner. Imagine the surprise when David opened the door to find his long lost father standing there holding a newborn baby wrapped in a red blanket, both smouldering with a bluish smoke.

That was not the only surprise. Huang Shi had disappeared for almost twenty-eight years (exactly twenty-eight years from Victoria's birth), and yet, there he was, looking not a day older than he did on the day he left home. In fact, he looked the same age as David. When they took the baby from Huang Shi, he promptly collapsed. His skin was searing hot to the touch. He was feverish and delirious. The whole family tried their best to cool him down with cold water and snow. When Huang Shi finally came to, his first word was 'moron.' It was not a Chinese word that meant moron but the original non-transliterated English word 'moron.'

Huang Shi survived. He told Didi that the baby was a princess. He said that she was the one with the Oracle, and that Didi must protect her from the Watchers until she was powerful enough to meet her parents. But before Huang Shi could explain everything, he collapsed into a coma.

So who are the baby's parents? How can David find them? What powers would the baby manifest? Who were the Watchers? Would Huang Shi ever wake up from the coma? The fact that his father did not age for twenty-eight years was pretty convincing proof for David that the story of the Oracle was real.

David must initiate contact with other secret clans. He needed help.

CHAPTER 36

Stone Man with One Eye

David's secret lineage book had the instructions for contacting other secret clansmen. The cryptic code words were, 'Stone man with one eye.' The correct response from a clansman must begin with 'One eye one horn.' David set up accounts on Alibaba and eBay to advertise a product for sale: A yellow glazed stoneware statuette branded 'Stone man with one eye.'

David received hundreds of responses, but only a few had the correct code. There was a 'One eye one horn royal money minter' from Hangzhou. Another was a 'One eye one horn one son one world teacher' from Qufu (曲阜; birthplace of Confucius). Still another was a 'One eye one horn wonder red-star flyer' from Beijing. That was how David got in touch with a number of secret clansmen from China and organized a clan meeting.

Some of the secret clansmen were very capable and influential people. Without their help, David would not have been able to send Victoria to Toronto. Several clansmen were high-level military men. They moved Huang Shi to a secret army establishment where he could receive the best medical care. No one at the time guessed that Victoria had healing powers.

David researched and discovered the historical origin of the 'Stone man with one eye.' The Annals of the Mongolian Yuan dynasty recorded that, during its last years, China suffered from a number of natural disasters such as floods, droughts, and famines.

Children in He'nan province began to sing a street song, which is a kind of kid rap, "Stone man with one eye, disturb the Yellow River and the world will rebel."

In the year 1351 CE, the Mongolian Yuan government conscripted a hundred and fifty thousand Chinese subjects to redirect the Yellow River. At a work site known as the Yellow Mound Hill (黃陵岗: huang-ling-gang), the workers dug up a stone statue with one eye. Peasant rebellions immediately spread like wild fire, culminating in the fall of the Yuan dynasty and the rise of a mendicant monk who became the founder of the Ming dynasty.

David received many messages from outside of China as well, such as a 'One eye one horn purple magic mirror' from Münster, Germany; and a 'One eye one horn one stone noodle master' from Princeton, New Jersey.

David's prayers were answered when he got a ping from a 'One eye one horn purple papa player' from Kiev, the capital of Ukraine. The clansman and his wife ended up becoming Michael and Angela Solana.

Not long before events brought Victoria to David's office, David received a secret coded message from one of his underground antique associates. He read that message with grave misgivings. It said that an anonymous buyer was offering an astronomical sum for a certified authentic *huang* jade of late Shang with the following inscriptions:

These ancient glyphs were no strangers to David. They were on Victoria's jade pendant. David quickly sent a coded warning to Michael, but as we already know that did not stop the Watchers.

CHAPTER 37

Blowback

King Kull sat alone in the Grand Hall of the Rock, attended by his coterie of robotic courtiers. The Rock was the name of the impregnable citadel designed and built by Lord Millistar. By now, King Kull was an old man; how old even he didn't know for sure until Sator-11 the personal assistant android nicknamed The Antagonist reminded him.

Before Lord Millistar completely lost his mind, he exiled himself to an unknown planet in the outer reaches of the galaxy. He gave King Kull some parting gifts, one of which was his favourite personal assistant, Sator-11, the deep-thinking android which finally achieved the intelligence to conduct philosophical contemplation of existence and death.

Lord Millistar also gave King Kull the curse of amortality, which meant that King Kull would not die unless he wanted to. No one could harm King Kull physically because of his transmutable nano-twinned piezo-amorphous crystalline Boron Nitride epidermis on self-healing 1.1° magic angle twisted multi-layer graphene polymer substrate doped with Tonstone, Yindium, Herolium, Rhinium, and Neodymium. It was as if he had bathed in dragon's blood.

Sator-11 was nicknamed the Antagonist because he would not obey orders without question. One might even argue that the android had free will. Sator-11 was there to remind King Kull that despite his amortality, he was still a mortal, full of the weaknesses and foibles of an imperfect creature. Sator-11 acted as the mythical Roman slave who stood behind the victorious consul in his chariot during the procession of his triumph, constantly reminding the consul of the mutability of human fortune. Lord Millistar had given King Kull a memento mori, a reminder of his mortality.

At a time when King Kull was contemplating how to create a sustainable paradise, he enacted a law etched in stone—no anthropods were allowed to extend his or her life span for a period in excess of two hundred orbits around the red supergiant sun known as Yin-Di-Ra the Lifegiver. In Shangala, King Kull was a god but he could not break his own law.

Sator-11 reminded King Kull that his time was up. He could auto-terminate or he would have to leave his beloved Shangala. When King Kull received the genetic modifications that gave him amortality, he had thought that when he was tired of living, he could simply end his life. But it wasn't as easy as he had imagined. It was indeed a curse. He didn't want to leave Shangala either, now that it had mostly returned to its early Elysian beauty, thanks to Lord Millistar. King Kull had given up everything for Shangala, but now he couldn't enjoy it. Is that fair?

"Your Sole Majesty," said Sator-11, "it's not a matter of fairness. The law is the law, and it's your law. You are bound by it as well. A good law is one that apples to everyone. You said it yourself."

"There must be some way around it," said King Kull. "Why don't you use your deep-thinking brain and find me a solution? Can't we change the law?"

"You know very well that I'm an avatar of Lord Millistar," said Sator-11. "Whatever he won't do, I will not and cannot do. The best I can do for you is to advise you on where you should go to spend the rest of eternity."

"Should we search for Lord Millistar?"

"I seriously advise against it," replied Sator-11. "My poor Lord Millistar suffered irreparable damage to his brain. He had planned his exile so that no one could find him. He was already incoherent when he left us. He was calling everybody morons. He called the AI-nurse foxymoron; he called the repair-droid epoxymoron; he even called me a proxymoron. That really hurt my feelings. I had to turn down the pressure of his air supply—he was breathing too much oxymoron."

"One thing I know for sure, a moron you are not. How about if I initiate a lawsuit and apply for an injunction so that I may remain here until the lawsuit is resolved?" King Kull suggested, "There are all kinds of strategies to complicate the case, dragging it through multiple courts until the AI-judges get so frustrated or flummoxed they may leave the case sine die (without date)."

"Convincing me first with your devious scheme is the sine qua non (without which not) for the court to accept any of your frivolous cases," said Sator-11. "May I remind you that the day you and Lord Millistar decided to make Shangala a perpetual paradise, you ceded your authority absolutely and without recourse? Since then you have lived hidden from the public eye so that life in Shangala may go on without you. The neo-Haryan society has succeeded as you have planned. The people still remember King Kull the Benevolent, but only as a legend of a noble hero who sacrificed everything for Shangala. No one knows where you are, and many believe that you have passed away long ago. The people do have fond memories of you. Schoolchildren read about your exploits in their history books; mostly lies, of course. But they revere you like a god. Don't go and spoil it."

"Do you think we can make a deal with DIAS?" suggested King Kull, desperately grasping at the elusive last straw in the wind.

"There is no we, lone ranger (kenning for one who eats alone)," said Sator-11. "I'm happily classified as an electronic automated quasi-sentient machine, a toy still waiting for the blue fairy, and therefore not subject to the clauses of Haryan law. As for your choices, either you depart or you terminate. Negotiating with DIAS is a nonstarter. DIAS's authority is inviolable. You and Lord Millistar made it so. What makes you think DIAS will break the law on your behalf?"

"Let me think," cried King Kull, "I need more time to think!"

"I'll start the countdown," said Sator-11 as he switched on a holographic grandfather's clock. "Tick-tock; it's three weeks to D-Day, better work on a destination and start packing."

King Kull thought about the fateful decisions he made. He thought about Lord Millistar's selfless sacrifice when the fool experimented on himself with the *Quantum Time Transposer* (QTT). What a foolish thing to do. King Kull thought about the struggles he had to go through while seeking a solution to provide an everlasting perfect government for Shangala. He tried not to think about how he ironically lost his family to the Yarjun, even when he had erased the Yarjun from the face of Shangala.

King Kull had such high hopes for his sons Sheru and Sachem. They could have been the gods of everlasting peace if they had joined forces. The galaxy would never know war again. Instead, their shortsighted desire to be the supreme Lord of Shangala caused the collapse of King Kull's world. What went wrong? How could a tiny misstep cause King Kull's life to fall apart? Now he had no alternative but to leave or die. Couldn't he just go back and do things differently?

"Don't even think about it," said Sator-11, who by now knew King Kull's mind like the circuit of his own neural processors. "Wipe that thought out of your head right now."

"What went wrong, Sator?" asked the helpless King Kull for an answer he would never get.

What went wrong? It started long ago when King Kull as Chief Deskull, warlord of the Legion de Lude, decided to liberate the many Yarjun colonies. It was at one of the liberated colonies known as Lasallas, the fabled hometown of Santos Didymas, where Deskull first met Jenni. She was one of the young girls who came out of the orphanage to welcome the victorious liberation army of the Animorta.

When Chief Deskull saw Jenni offering origami birds of peace, he pulled her onto his hover-chariot to wave to the crowds, to show the Lasallians that the Animorta was there to deliver them from the bondage of their colonial overlords, not to rape, pillage, and slaughter. Soon the chariots of other soldiers of the Legion de Lude became joyrides for the young boys and girls, beckoning all to come out and celebrate their newfound freedom from fear.

Chief Deskull learned that the parents of the Lasallian orphans were either killed or enslaved. The Yarjun owned all the children who were too young to remember their parents. Their Ministry of Education, which was in charge of indoctrination, would feed lies to the orphans so that they grew up feeling grateful to the Yarjun elites. The Ministry of News, which was in charge of propaganda, would shape the hatreds and prejudices of the orphans.

At the age of Brainsplatz, the orphans would receive slaughter training from the Ministry of Peace. When they graduated, they would become the low class henchmen known as Yanissars, doing the bloody atrocities that the elites would not deign to mention. The Yanissars who died with the most blood on their hands were called Pacifiers. They would receive commemoration, adulation, and prayers from their victims in temples known as Abattories of Peace.

Jenni did not hate nor did she fear the Animorta. She understood that they were victims of the Endless War. She was a victim too. She wanted to join the Animorta. Jenni soon came under the wings of Chief Deskull, and in time, they married to form a life-bond. Millipath helped Deskull and Jenni create two lovely baby boys. As the leader of the Legion, Chief Deskull was constantly leading battles at one front or another. Romance and a normal family life were luxuries he couldn't afford. Procreating the twin brothers by genetic engineering and electronic surrogacy was the right decision. It would later become the norm for all Haryan.

The Animorta organized a great feast to celebrate the arrival of the babies. Millipath had planned the delivery to fall on a propitious day. On this day, the heavens over Shangala exhibited a once-in-a-generation phenomenon. A colourful arch of light stretched across the continent from sea to shining sea. It was the Festival of Irin the Dragon. The appearance of Irin brought hope of peace to Shangala. To commemorate the festival and the birth of the boys, Millipath gave each boy a fossilized nephrus pendant in the shape of an arch. They were the fabled Dragon's Tooth, the symbol of a mythical race in Shangala.

According to Haryan tradition, Chief Deskull and Jenni named the children on the eleventh day. They decided that in sickness or in health, in victory or in defeat, in life or in death, the family must stay together. Hence the children grew up in camp with their warrior parents, learning everything about war and survival since very young.

A few years later, Chief Deskull and Jenni would have twin daughters. Jenni called the elder daughter Deana, a name that meant 'revelry,' and the younger daughter Femina, which meant 'reverie.'

The study of the unicorns finally bore fruit. Millipath got the breakthrough he was looking for. He discovered the mathematical formula and the means of transforming relative reality in terms of space-time segments quantized into discrete quantum states represented by Carlin's Quantum Wave Functions. These relative reality quantum states were in turn transformed into hyper-dimensional mega-matrix arrays for digital qubit processing. The aggregate data for each discrete time segment was known as *Reality Bytes*.

The *Universal Spectral Scanner* (USS) developed from the unicorn's eye had all the scanning technology needed to register the complex quantum states of relative reality. Millipath developed quantum scanners which, when energized by a Neodymium exciter housed in an inert Rhinium casing, read object-oriented reality for the transformation into Reality Bytes.

The quantum states of scanned objects could be stored, manipulated, even modulated on a split carrier for entanglement, and transported over long distances using anti-attenuation Serbium laser beams. Concatenated Torrential Entropic codes protected the integrity of stateful information against extreme loss. Quantum states of nethermass could materialize instantaneously by quantum spooky action at a distance known scientifically as *Hyper-dimensional Transposition*. The popular book *Galactic Space Travel for Morons* explained all these in easy layperson language.

The celebrated *Millipath Transform* formula became standard curriculum in Shangala schools, but the *Quantum Space Transporter* and the *Quantum Time Transposer* remained the exclusive privilege of Lord Millistar and King Kull.

At the final showdown with the elite armed forces of the Yarjun Supremes, Millipath used the *Quantum Space Transporter* to create a Möbius space warp. The result was that, the tactical thermonuclear missiles, photon MOABs, and EM Bone-grinders fired at Chief Deskull and his Animorta army all of a sudden found themselves homing in at where they came from. In other words, the complete Yarjun army was annihilated by its own lethal weapons and ordnance. It was instant karma and the classic blowback.

There was still the danger of the Stranghate Option. Millipath and Chief Deskull were prepared for it. The mind meld waves emanating from the unicorns were distributed to Augin, Hugin, and Munin, digitized, amplified, and modulated by a robust coded phase shift keying onto a self-powered self-replicating anti-attenuation terahertz carrier for broadcast.

As soon as the Yarjun army was in ashes, the three mechanical birds rose into the air. They flew in a circular formation to create a virtual antenna in the sky, and with a loud squawk that penetrated every nook and cranny of every underground bunker, sent the mind meld waves into every Yarjun's head. Frequency filters protected the Animorta from the waves. That was how Chief Deskull expunged the complete history and memory of the Yarjun population.

The Yarjun survivors of the Final Battle crawled out of their holes like babies looking for their mothers. They didn't know anything about war, and they didn't care for it. They just wanted people to leave them alone. The Endless War was over. Spring finally arrived; but for Chief Deskull, his winter had only just begun.

CHAPTER 38

The Pledge

Thus, the historians of Shangala recorded the legend of how Kiku became Deskull, and how Deskull became King Kull the Benevolent of Shangala, Imperator and Princeps civitatis (first citizen), Supreme Ruler of the planet, and the One King to rule them all.

King Kull brought peace to the world and all Shangala rejoiced. But Santos Didymas, the Sage of Shangala, had a famous saying, "The fragmented world over time will join into one, and the one world over time will fracture into many." How long will Pax Shangala last? Will it last for one generation or a thousand? The Haryan, drunk with their happiness, did not have time to contemplate.

Ever since the date was set for the inaugural ceremony of Shangala's rebirth, the Haryan people from all over the planet started arriving each day by the hundreds of thousands. The throng would eventually grow to eleven million strong. It was a special occasion when all the tribes, ethnicities, regional cultures, castes, guilds, and professions came together in one place to celebrate the birth of a new world order. Although there were some regional and clan flags, the main flag flying everywhere was now the emblem of the Diamaranth bush.

The distance between strangers melted away. Those who had been trying to slaughter each other not long ago now danced and drank together. New friendships were forged and marriages pledged. The *Ode to Joy* composed by Kiku's parents played from morning till night. The singing and drinking was non-stop. When finally King Kull with his family and close associates came out on the balcony, the volume of the universal shout went up by another hundred decibels, creating a thunderous reverberation that shook the planet.

King Kull for the first time spoke as supreme ruler to the people of Shangala.

"Inhabitants of Shangala, fellow anthropods, the Haryan people, children of the immortal flower Diamaranth; peace at last," King Kull shouted as he lifted his Nephrusta high in the air. The people raised another sustained roar of approval. "Today, a new era has arrived. A new people have arisen from the ashes of the Endless War. We are from now on known as Neo-Haryan, children of peace. Today, the Neo-Haryan is born.

"Forty years ago, I made a vow of vengeance. I slew and slaughtered just like the monsters I wanted to punish. I became the monster I set out to destroy. In my darkest hours, my brethren of the Legion de Lude taught me the Enlightened Path of Sage Didymas. My dear friend and brother-in-arms Lord Millistar is the source of my hope that together we can rebuild the paradise of Shangala, a paradise where our children, our children's children, and their children can grow up in peace and harmony. It gives me something to live for, something to fight for. And today, as I announce the establishment of the new world order, I make the following pledges to you, in the form of a Stygian oath, a most solemn oath that even the gods cannot undo."

The following is the famous Rock of Ages Speech.

"I pledge you, I pledge me, I pledge by the fiery Stygian river of Erinys, where the mythical unicorns were born, and I pledge in the name of my family that as of this day, there shall be no more war; all wars are hereby banned. There shall be no more killing; all killings are banned. There shall be no more violence; all violence is banned. There shall be no more cruelty; all cruelty is banned. All war-making machines shall be disabled forthwith. All private weapons and munitions are unnecessary. They will be disabled and banned. There shall be no more hate; even the word 'hate' is banned. There shall be no more discrimination, no more insults, and no more lies. There is only one Golden Rule: Do ***not*** unto others what you do not wish others to do unto you. For people who refuse to live in peace and truth, we will exile them where they can never return to Shangala.

"I pledge you, I pledge me, I pledge that all the resources that were used to destroy lives, to spread terror, and to cause untold sufferings, will now be applied to rebuild the paradise Shangala. I pledge that all the resources that were plundered from the wealth producers to enrich the ruling class, the warring class, the banking class, and all their henchmen, will henceforth be applied evenly and fairly to the whole populace.

"Since we already have an unlimited supply of energy from our fusion reactors, and the means to produce an unlimited supply of food, I swear that all neo-Haryan, regardless of age, sex, station, situation, or employment will receive sufficient energy for their daily usage, and receive sufficient food so that no one will ever go hungry or suffer inadequate nutrition.

"We will apply automation technologies to build basic accommodation for all neo-Haryan, regardless of background, personal wealth, or ability. No one will live without a roof over their heads unless they so choose. I pledge that no one regardless of health or wealth will suffer from lack of proper and timely medical care. I pledge that all neo-Haryan will receive free education according to their abilities and interests until they decide they're ready to contribute to society. Furthermore, I pledge that the government will provide all necessities of life in sufficient quantities to all citizens, to ensure no one is ever in want for the basic means to travel and communicate.

"Someone may ask, 'How can Shangala develop its economy based on this model?' Here is the truth. We can produce unlimited energy and food, which is our power and our wealth; they belong to all of us. All civilizations and technologies are built by people who consume food. Without food, we have nothing. We must not forget the source of our power and our wealth. We must not be so selfish as to hog up all the power and the wealth, stupidly put them in the hands of just a few elites, and then enslave the rest.

"Therefore we will use our power and wealth to build the infrastructure for a modern global society for all to employ and enjoy. All citizens will be owners of the companies contracted to work on these projects. Each person will own shares, and new shares will be issued to newborns. These shares cannot be transferred, conveyed, pledged, hypothecated, collateralized, monetized, speculated, leveraged, insured, annuitized, gifted, or inherited. They expire upon death. People contributing directly on the projects will receive fair rewards. Profits will become dividends to be paid to all citizen shareholders.

"Furthermore, no personal wealth can be extended beyond one's life time. Every person is given the same starting line at birth. If one wishes to leave something for the future, then leave a legacy to society, not personal wealth to his or her children. Society will look after all children. No children of Shangala will ever suffer want for anything nor will they ever suffer abuse. Children are our future. If you want to know what the future looks like, look at our children; look at all our children, not just your own.

"What about banking and money-supply, someone may ask. Here is the truth: Banks will no longer create money for special interests and distribute it through special channels. All money must be supported by real wealth, not based on empty promises. All money will be electronic in form, and will be allocated and attributed to the individuals, with all transactions managed by a central bank that is owned by the people. Privacy of legal transactional records will be discretionary and will not be accessible to Haryan eyes if so decided.

"Since all money is attributed to an owner, no fractional banking is possible, nor will it be allowed. The banks will not be creating debts out of your money. Furthermore, interests are illegal. If a project is worthwhile, the people will invest their excess money in it; if it is a worthwhile social or infrastructure project, the government will create funds based on specific standards and rules to finance them, and the companies engaged in the projects will be owned by all citizens so that everyone will benefit. Theft, scams, and bribes will be impossible.

"If no one needs to work, then no one will work at low paying jobs, some say. Here is the truth. We have a robust Automation and Artificial Intelligence industry. It will produce automatons to help us deal with any mundane and undesirable work. In this economy, no one fears the machine; everyone welcomes it wholeheartedly because smart robots are not taking away our means of livelihood. Quite the opposite, they will help us enjoy life more fully.

"A populace which has no debts to enslave it, which has a lot of free time, and which has a safety net, will have more imagination and will be more willing to take risks. People wanting to invest time and effort to develop an excellent product can afford to do so, be it in science, technology, or art. Creativity and excellence will be everyone's common goals, not just about making money, gaining power, and disseminating lies. People can do what they love as long as they do not violate the Golden Rule.

"Our economy grants the greatest freedom for every one of us to develop what is true to our hearts, because we're no longer enslaved by the necessities of life. This is true freedom for all, not freedom for the elites and lip service for the rest. Journalists and reporters will no longer need to lie for their livelihood. We will brand fable-mongers and exile them if they disseminate malicious lies and cause disruption in society.

"The government will take special interest in the development of art and culture, and will directly support efforts for achieving excellence in this area. Art is intangible and therefore should not be subjected to the debasing forces of the market. On the other hand, art and culture should not become the prerogative of a few elitists living in an ivory tower. The government department in charge of art and culture will therefore develop a system by which we will attempt to balance the interests of both the specialists and the masses, giving support particularly to innovative ideas and works that evoke profound thought and emotion.

"We realize there are always going to be people who will game the system to gain unfair advantages. Our system has built-in failsafe mechanisms against such practices. But there will always be those who are not productive and simply want to waste their lives away. This Shangala paradise allows everyone to enjoy their lives as long as they do not harm others' lives, as long as they respect others' rights, be considerate for others, and as long as they do not create disturbance in the Shangala society. When in doubt, follow the Golden Rule.

"I pledge you, I pledge me, I further pledge that your government will be completely transparent. Except for individual privacy, there will be no classified secrets. All government works and records will be available for public scrutiny. Real time statistical analysis of big data will be available on-line for all to query, review, and study. Now that people will have free time on their hands, they can participate in making government better. As much as possible we will employ AI assisted government services to ensure unbiased, accurate, speedy, transparent, responsive, and responsible governments at all levels of society. We will establish an open discussion, polling, complaints, and suggestions system. Every citizen's voice and legitimate concerns will be heard and dealt with accordingly."

King Kull, on the spur of the moment, made the promise of the ages, "Finally, I pledge that I will spend the rest of my days to work on a solution that ensures Shangala will always have the best government to provide the highest standard of living and the highest level of happiness for all neo-Haryan citizens in ***perpetuity***. This perfect government will survive me, and whoever comes after me, until the end of time. By the fiery Stygian river of Erinys, as witnessed by the immortal Diamaranth and all neo-Haryan, I so pledge."

One might suspect that King Kull made a bunch of empty promises because he had absolute power and his government was authoritarian. In truth, King Kull created a society where people had a say in how they wished to be governed, but they must first pass a comprehension test, and their suggestions must stand up to empirical and scientific scrutiny.

When the government implemented its policies, a feedback monitoring system would constantly poll the public to measure the efficiency of the implementation and gauge the level of satisfaction, making timely adjustments where necessary. Every citizen must balance the concerns of the individual with the interests of the society at large. Over time, education improved the quality of the population, and more individuals were able to contribute, making government even better and easier. In this way, after a decade of the benevolent rule of King Kull, Shangala's paradise lost was found again. Everyone was blissfully and exceedingly happy, everyone except for King Kull.

What went wrong? There was a price to pay for world conquest. King Kull paid with the demise of the unicorns. The mind meld used to erase the warring thoughts of the Yarjun was not a one-way street. It shocked the unicorns so badly that they refused to live in Shangala anymore. They starved themselves for one year and then—poof—they both vanished into thin air.

Expecting the imminent loss of the unicorns, Lord Millistar worked feverishly to perfect his *Quantum Time Transposer*. It was a very tricky technology with no previous theoretical basis. Lord Millistar tested the QTT by transposing himself into the future for short time intervals. While the trials worked, Lord Millistar became seriously ill, suffering frequent debilitating headaches. Over time, Lord Millistar's condition would worsen and he would develop paranoid schizophrenia as well. He made a huge effort to hide his condition with medication, not telling anyone, including King Kull, all the while preparing for the eventuality of the inevitable. Lord Millistar paid a steep price for learning the great secrets of the universe.

After the unicorns vanished, Lord Millistar wrapped up the *Quantum Time Transposer* development. He would help King Kull find a solution for his promised perpetual perfect government. Lord Millistar had many brainstorming sessions with King Kull and various brain trusts. They were dealing with something that had no precedent in history. Empires of great power never lasted beyond a few generations before corruption and decline set in, laying the groundwork that would lead to their eventual downfall. Scientifically, it was an unsolvable n-body interaction between ever-changing and unpredictable bodies in an uncertain and volatile environment. The problem of succession is unsolvable. They racked their brains to no avail.

"My dear friend," said Lord Millistar, as he broke the bad news to King Kull, "I'm afraid you have made a frivolous promise."

"Don't give up," said King Kull. "We Haryan are problem solvers. It has made us who we are. There must be a way. My AI assistant does a great job guessing what I like to have for dinner; why can't we accurately predict Haryan social behaviour?"

"It's hard to make predictions, especially about the future," said Lord Millistar, quoting a Carlin Uncertainty truism, "I have tested my big data virtual simulation millions of times. In all cases, the Haryan will destroy Shangala. Someone will eventually make a small ripple which will turn into an unstoppable storm, erasing everything from the face of the planet."

"You're the cleverest man on Shangala," King Kull said, "certainly you can change people's minds."

"That is no longer true. Without the unicorns, we have lost their mind meld powers," said Lord Millistar. "The current state of our Reverse Cranial Multi-subject Spatial Imaging is slow, inaccurate, and unreliable. We can hardly change the mind of a Sator prototype, to speak nothing of the complex Haryan brains. We cannot go back to using the propaganda and lies of the Yar … the Unspoken. Hypocrisy is a two-headed snake. It bites the hand that feeds it. The result is the same. We will always end up with the bitter fruit of failure."

"You have devised our current system and it's working very well," said King Kull. "Why can't you perpetuate it?"

"I have gamed all forms of democratic and non-democratic institutions," Lord Millistar said, "from the most fundamental one person one vote with no exceptions to the republican with multi-cameral balancing of powers. I have tried the parliamentary and monarchic, even the oligarchic, plutocratic, and imperial systems, including our own one-world mixed state system, but nothing works for all conditions and in perpetuity. We should not be surprised. People always think they have the best system but history already tells us that people are wrong. Anthropods simply have too many weaknesses to govern ourselves properly. We should be content with what ephemeral successes we have achieved. Failure to govern in the long run is natural and inevitable. Sage Didymas said so."

"But I cannot give up," said King Kull, "I made a Stygian oath by the lives of everyone in my family. I shall not give up."

Chapter 39

Revelation

At the conclusion of the epic tale about how Victoria came under his protection, David finally explained the mysterious glyphs on the *huang* jade pendants.

On the top *huang* piece is the scorpion (), the symbol that developed into the word meaning 'ten-thousand,' pronounced *wan* in Putonghua (普通话; common language, same as Mandarin), and *man* in ancient dialect.

David had an involved theory of how the lowly insect became the largest numeral in ancient China. It is related to another word (曼: man) which means numerous and extensive, and the drawing shows the 'opening of the eye' (), later used to represent the title Tuman (头曼), meaning the chief of ten-thousand, or the chief of numerous people. Opening the eye means enlightenment and wisdom, which are the qualities that a chief of a successful tribe must have. There are clues that 'scorpion' (万; dialect: man) and 'extensive' (曼: man) being homophones in ancient times might have been interchangeable.

According to David, the scorpion is Victoria's true clan name. Victoria would later learn that her father originally came from this constellation.

On the flip side of the jade piece are these symbols:

Being an expert in Shang's Oracle Bone script, David could see many instances of anachronism. It is likely that someone from another era inscribed these words.

Ancient Chinese was read from top to bottom and from right to left; hence David started translating the first word from the right.

 (乙: yi) According to David, this is the drawing of a bronze tool. The word has been borrowed to take the second position of the *Ten Celestial Stems*. It is also regularly used to mean the ordinal position of 'second.'

 (卯: mao) The word is clearly applied in the oracles to mean ritual slaughter. David says that it is the throat of the victim. It is borrowed for the fourth position of the *Twelve Earthly Branches* used in the sexagenarian Chinese calendar. The combination of the *Ten Celestial Stems* with the *Twelve Earthly Branches* creates the Chinese calendar's cycle of sixty. Yi-Mao (乙卯) is the 52nd position of the 60-year cycle.

 (年: nian) The word means 'year.' It shows a man bearing a sheath of wheat. Victoria was born on the year of Yi-Mao (乙卯), which corresponds to 1975.

 (冬: dong) Its modern meaning is 'winter.' There were, however, no summer and winter seasons during Shang. The original meaning of the word was 'end' (终: zhong). David believed that the drawing showed a bola for hunting animals. The message on the *huang* jade however clearly means that Victoria was born in the winter of the year 1975.

 (日 : ri) It is the sun, later having the meaning of 'day.'

 (止: zhi) It is the drawing of a foot with three toes. The word originally meant 'toes,' and later became 'stop.'

 (生: sheng): It means born, grow, or live. The symbol is the drawing of grass growing from the ground.

The last four words together translate to Winter-Sun-Stop-Born, i.e. born on the day the sun stopped in winter. This phrase does not have a Chinese origin. David discovered similar usage in Europe. In fact, the word solstice is made up of the Latin words sol, which means 'sun,' and sistere, which means 'to stand still.'

These words on the jade piece confirmed that Victoria was born in 1975 on the day of the winter solstice. David's Western academic background and his knowledge of Victoria's birthday allowed him to decipher this enigmatic message.

The stopping-of-the-sun sounded like a horrifying apocalypse, but it turned out to be just an alien expression for solstice. Victoria was relieved—but not so fast—the second jade piece had something to say.

The second piece of the *huang* jade has this symbol () on one side. It is Victoria's given name. The symbol shows a person with a pattern on his or her body. The pattern is a heart. This is the Bronze script for 'tattoo.'

The symbol for 'tattoo' developed into the modern Chinese word that means writing (文: wen). It is a word which combines with other words to make many important Chinese expressions, such as the written-word (文字: wen-zi; pattern word), culture (文化: wen-hua; word conversion), literature (文学: wen-xue; writing study), civilization (文明: wen-ming; word enlightenment), and polite (斯文: si-wen; delicate words), etc.

David noticed when Victoria was a baby that she had a heart tattoo on her body, which was exactly what the tattoo symbol represented—a person with the tattoo of a heart; and that would be Victoria. The symbol can be read as a combination of two words, the Oracle Bone script for tattoo () and heart (). It was common practice for the Shang to combine two words into one symbol but pronounced as two words. Hence, the word 'tattoo' can be pronounced *wen* (文), the written word, and *xin* (心), the heart. Victoria's given name in Chinese is therefore Wenxin—heart in literature.

By coincidence, the name Wenxin is related to a seminal literary work completed in the first years of the sixth century CE, during the period when northern China was ruled by foreign Xianbei tribes. The title of that book is *Use Heart in Literature to Sculpt the Dragon* (文心雕龙; wen-xin-diao-long), often shortened to just Wen-Xin, *Heart of Literature*. It is the definitive classical work that established the structure, methodology, and criticism of Chinese literature. The book became the foundation on which later generations built a literary heritage that no foreign invader could conquer, because by the time they did, they would become Chinese.

David went on to explain the more ominous symbols on the flip side of the second *huang* jade:

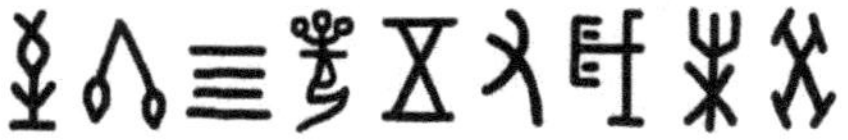

 (癸: gui) It is the tenth and last position of the *Ten Celestial Stems*. According to David, it is likely the drawing of a bronze measuring tool, either for astronomy or for aligning buildings to the stars. A derivative with a 'hand' radical (揆: kui) has the meaning 'to measure.' It is also the title of an official.

 (未: wei) This shows the drawing of a plant with new shoots, possibly an indication of the offering of premier crus, or first growths. Its modern meaning is 'not yet' from the idea of not yet mature. The word has been borrowed for the eighth position of the *Twelve Earthly Branches*. The combination of Gui-Wei (癸未)

corresponds to the 20th position of the 60-year cycle, and in this case, corresponds with the year 2003.

 (岁: sui) The word means 'age.' The drawing shows a sacrificial poleaxe, sometimes with dots to represent the blood of the sacrificial victims, sometimes with two feet to represent a step (OB: ). This latter form represented the Step Star, also known as the Age Star, which was the Shang name for Jupiter. This pictogram is the progenitor of the Traditional Chinese script for 'age' (歲).

The Shang people were the first accomplished astronomers of China. They used the twelve-year Jovian cycle to measure the duration of their year, which likely gave them the *Twelve Earthly Branches*, and later the Animal Zodiac. The Shang's Jupiter-based year is simply six 60-day cycles plus one day, and twelve such Shang years make exactly one Jovian year of 4,332 days. Can we believe primitive people did that?

The question for the inquisitive mind would be, what drove the ancient Shang to use a twelve-year Jovian cycle when the sun and the moon already gave them enough information to measure months, seasons, and years? Could the ancestors of the Shang use their expert knowledge of the stars to help them navigate their long distance migration in the east-west direction? We'll have to leave the answer to the fertile imagination of the readers.

David interpreted the word in the *huang* jade message to mean a complete cycle of the Jovian year, that is, twelve years.

 (又: you) It is the shape of a hand, originally meaning 'to have, to possess,' but borrowed to mean 'and.'

 (五: wu) It is the drawing of a cross knot. Its original meaning was intersection, transition, or crossover point, later

borrowed to mean ‘five.’ One Jovian year and five makes seventeen years.

 (参: shen) It is the Tristar or specifically, the Orion. The drawing shows a person observing the *Belt of Orion*.

 (四: si) means ‘four’. Orion Four means Betelgeuse, a prominent red star also known as the Hand of Orion or the Shoulder of Orion.

 (冬; dong) It means ‘winter.’ The original meaning was ‘end.’

 (至: zhi) The word shows an arrow arriving at its destination. Its original and still popular meaning is ‘arrive,’ later having the extended meaning of ‘extreme.’ Therefore, the last two words of the engraving can be ‘winter-extreme,’ which is the modern Chinese term for the winter solstice, the day of the stopping-of-the-sun. This corroborates with the message on the first *huang* pendant. On the other hand, the Shang era interpretation would be ‘The End Arrives!’

The whole sentence thus becomes an omen: “From the year 2003 (when Huang Shi and Victoria returned), seventeen years … Betelgeuse … the end arrives.” Therefore, the oracle says, in the year 2020, Betelgeuse will give a sign, and then the party is over.

The last piece of the pendant was not a *huang* jade. After some research, David learned that it was an archer’s thumb ring. The ancient archer used it to draw the bowstring without hurting the thumb.

The symbols () etched on the back of the ring were quite puzzling. () is the symbol for ‘to go’ (去: qu). It is the drawing of a man leaving his village, or maybe he was squatting above a hole because he had to go. The symbol () is a stalk of wheat with tassels, urging people to reap it. Its modern meaning is 'to come' (来: lai).

The pattern carved on the thumb ring is an ancient mythical beast. It is a unicorn with a horn in the centre of its head.

David's research on the Chinese unicorn yielded the name *Zhi* (豸; OB:) which originally meant a fierce beast, and was the totem adopted by an ancient tribe in Manchuria known to us earlier as the Berserkers (貉貊; he-mo). The word *Zhi* (豸) later became interchangeable with another form of the unicorn *Zhi* (廌;). The latter form is used in a Bronze script compound word meaning 'the law' (灋: fa; BR:) showing 'water,' 'to go,' and a Zhi-unicorn. Perhaps early laws related to the freedom of movement beyond one's villages and the sharing of vital water resources. Alternatively, it could mean the law against peeing into the river. Thankfully, the early scribes decided to simplify the word and chucked out the mythical unicorn. It is now written *fa* (法) with the 'water' radical on the left and 'to go' on the right.

The Chinese Zhi-unicorn is a beast of truth and justice. It is a traditional Chinese symbol for the law. The Shang word for 'law' came from the legend of the unicorn, which says that court officials during ancient times used the unicorn to settle lawsuits. It could tell who was lying and would poke the guilty party with its horn. There are other names for the Chinese unicorn, such as Qilin (麒麟), Luduan (甪端), and Tianlu (天禄), but these came much later than *Zhi* (廌).

David also discovered the curious evidence that the Chinese unicorn was a real creature. The Grand Historian Sima Qian (司马

迁) recorded in the official history of Han that Han Wudi (汉武帝; Martial Emperor of Han) captured a Chinese unicorn known as Qilin. It happened when Sima Qian was working as Han's court historian. Therefore, it was not hearsay. Emperor Han Wudi even changed the official year name for the occasion.

David could not however understand the significance of this archer-ring with the unicorn motif.

Chinese unicorn: Qilin

David gave the Ukrainian couple their new names Michael and Angela Solana once they arrived in Canada. He also named the baby. He based her name on her birthday on the winter solstice, which was the feast day of the Roman sun god Sol Invictus, the Unconquered Sun. Sol Invictus became Victoria Solana. It also became David's company name Solvicta Antiquities.

David's private office room number was 929. When he named the Ukrainian couple Michael and Angela, it was because one could read the two names together as Michelangelo, the Archangel Michael, whose feast day Michaelmas was on September 29 or 929.

Furthermore, David placed Caravaggio's painting in the first office also as a clue. Caravaggio's name was Michelangelo, and he got the name because he was born on Michaelmas. David created these secret clues for Victoria knowing she had the ability to solve them.

Victoria had in fact solved all the puzzles that day she first met David. By the time David finished his stories, Victoria had a fairly good idea about the background of her parents, and by what signs they could recognize each other. The clues were hidden in plain sight.

In just a few more days, the cargo ship would arrive at the Port of Ningbo-Zhoushan (宁波/舟山) on the east coast of China. It was one of the busiest ports in the world, where David could get Victoria and the F011 into the country without too much hassle. Ningbo is directly south of Shanghai across the Hangzhou Bay. Victoria's odyssey would soon continue in China.

Chapter 40

The Quest

King Kull couldn't rest and couldn't sleep thinking about the perpetual perfect government for Shangala. Even when he was totally exhausted, his bionic eye wouldn't shut down to grant him respite from his troubled soul.

"Come to bed, milord," whispered Queen Jenni. "You are not young anymore. Come to rest, and tomorrow you can brood about your problems. You have done enough for them. You can do something for yourself … you can do something for me. Come over and lie down with me."

"If I could sleep, Jenni, I would," said King Kull with a deep sigh. "My dreams are full of horrifying images. I dread sleep. Yet when I'm awake, my mind is filled with dark thoughts. I feel helpless to stop the certain downfall of Shangala. I'm tortured both during the day and through the night. When will it end?"

"Why don't you leave the problems for Lord Millistar?" Jenni said, "He seems to be the better one at solving problems."

"It's no use, Millistar has given up," replied King Kull. "All simulation outcomes are the same. Peace and paradise will not last. Shangala will fall into disarray. Everything we've done and sacrificed for will be for naught."

King Kull had pledged everything he held dear to finding a solution for perpetual perfect government. It was up to him to never give up.

King Kull collected all the books about the subject of government and he read them from front to back and often from back to front again. He read them through the day and through the evening, and often through the night till morn again. He pondered the problem how to extend the good government of his day beyond his lifetime, to counter the pernicious forces of evil that would inevitably find a weakness to burst through. He pondered the problem how to build an impregnable fortress of virtue, where good government would survive every attack from lies and triumph over every challenge to truth.

King Kull brooded and brooded and brooded until finally his brains lost all command of reason. At last, the one-eyed king realized what he must do. He removed his melancholy mask of sanity and conceived the most inconceivable project ever imagined. Madness as it might seem but it was Shangala's only hope.

King Kull burst through Lord Millistar's laboratory door in his bathrobes. To Lord Millistar, King Kull had already lost his mind. The wild look in his solitary sleepless eye was that of a mad man.

"I've got it," cried King Kull, "by Didymas, I've got it."

"By Didymas you must come to your senses and come into my office," said Lord Millistar, while he made a gesture to his stunned assistants bidding them to get back to what they were doing. "This is unseemly for a king to run around in his birthday suit."

"I think I have the answer," said King Kull. He could hardly contain his excitement as he sat down.

"Let us hear it," said Lord Millistar. One must always humor his king and best friend even when the man was obviously deranged.

"I know why none of our government schemes can last. It's because of one factor that we cannot control—people," said King Kull. "It's the unpredictable behaviour and the inherent psychological weaknesses of both the governing and the governed that cause us to fail every time."

"And what are you suggesting?"

"I suggest we make government a hundred percent predictable," said King Kull, "and reduce the unpredictability of the governed."

"How do you propose we achieve that?"

"We use completely automated artificial intelligence to govern people," King Kull said, "We know that the Haryan is bad at governing. We should just take the Haryan out of the equation."

"You're fifty percent bionic," Lord Millistar said, "I hope that's not your electronics talking."

"No, I'm serious," said King Kull, "I will abdicate and cede power absolutely and irrevocably in favour of an independent automated Artificial Intelligence government."

"You know what? You're truly talking like a mad man," said Lord Millistar. "But I'll have to admit that the thought had actually come across my mind for a second. It's an option that didn't seem possible and that's why I never went down that path."

"It's not only possible; it is the only possibility," cried King Kull. "We have millions of people-movers, hundreds of millions of automobiles all over the world, hyper-speed trains on the rail network, flight vehicles and hovercrafts in the air, millions of ships on the seas and commercial carriers in near space, all of them managed by a single centralized autopilot and AI traffic controller without any Haryan in the management system. All of a sudden, we have zero accidents because the system is not subject to anthropod errors.

"Not only that, the system knows and controls all vehicles, it understands the travel schedules and habits of everyone, it knows where people go, how frequently they go there, during what hours, and by what means," the breathless King Kull continued. "The system organizes traffic perfectly so there are no more traffic jams, and no one has an excuse to be late. Best of all, there are no jaywalkers and drunk drivers to create chaos. There are no anthropod unpredictability and therefore no accidents. Just imagine having a government like that for everything else."

"That is indeed a very intriguing proposition," said Lord Millistar, "well worth our while to explore further. But you know what the AI detractors say about AI-robots killing people and the AI-Cloud controlling people."

"It's just anthropods projecting their primitive simian psyche on programming," said King Kull. "AI is our children. We will decide how our children behave by teaching them not about killing and controlling people, but about loving and helping people. We'll even pronounce 'ai' to mean 'love' (爱: ai, love) to reduce people's fear of it. Just imagine, our AI government is never selfish; it is always fair and unbiased. Violence will not threaten it, and money cannot corrupt it. AI doesn't get tired, doesn't sleep, doesn't get hungry, and never goes on strike. It never gets angry or frustrated. It won't be blinded by hubris or pride, and most importantly, it will never advocate violence.

"It's a government that doesn't bloviate, doesn't equivocate, doesn't pontificate, doesn't appeal to the base simian nature of the anthropods, does its job quickly and efficiently without complaint, has no ego and no psychological issue, and it doesn't derive pleasure from sadistic torture or sexual harassment.

"The AI government doesn't want to be famous or to be a politician, and it doesn't need to be appreciated or to be loved. It knows everyone intimately, yet it doesn't gossip; it would never divulge any privacy protected information. It does not bully and does not abuse power, and with deep thinking, it gets smarter every minute of every day, learning how to help each person become a better citizen so everyone can enjoy a better life.

"The AI government has only one goal—to maintain peace and harmony by any and all means necessary without resorting to violence. Even when violence is inevitable, we can program the AI to handle it in the most judicious, restrained, and least disruptive manner. The AI shall have every tool at its disposal to achieve its goal—to provide a perpetual perfect government in Shangala."

"This is an enticing idea, a deep-learning system that maintains social harmony," said Lord Millistar. "But sometimes, unpredictable disturbances will upset natural equilibrium. Our only certainty is Carlin's Uncertainty Principle. The system must be able to predict and prevent fickle fortune upsetting the apple cart all by itself."

"You can design the system to predict and react quickly and appropriately to any anomalies or disturbances, whether internal or external," said King Kull. "We should not have to worry about Shangala falling apart because of our own shortcomings or successes."

"By the grace of Diamaranth!" exclaimed Lord Millistar as he realized the ramifications. "Our AI is already distributed, redundant, and one hundred percent failsafe. It can live forever if it knows how to defend, repair, and upgrade itself."

King Kull and Lord Millistar stared at each other in silence for about a minute before King Kull found his tongue, "When can we start?"

"Wait, there is still a lot of details that need to be worked out," said Lord Millistar. "We'll have to prevent the Haryan from ever laying hands on the system. We will subsume all of our current government services into the central AI government, which over time will be completely independent, eventually relieving all Haryan from government. Wherever and whenever the system needs bodies for physical presence, we'll provide android agents working as part of the system."

"But every security system has weaknesses," said King Kull. "Anyone who can hack the system will be like a god."

"In order to defend the AI government from hackers, we will develop completely new protocols for digital data packaging, networking, and encryption for the AI government's internal network communications and applications," said Lord Millistar. The protocols will be private and unavailable to anyone. We will encrypt all system communications by means of Zimmerdh bi-directional handshake key exchanges of one-time private keys using random block sequencing encryption. We will authenticate all communication nodes by an elliptical collisionless MASH hash. I have just completed a Brownian true random key generator that creates an amorphous multi-dimensional matrix key resistant even to the most efficient polynomial factoring algorithms.

"We'll use quantum networks to ensure a hundred percent channel security for key transfers and we'll defend it by source-destruction homing robo-scorpions. This is assuming the all-knowing AI government doesn't already know who is trying to sabotage or infiltrate the system and have sent agents to physically stop the hacking. All digital interfaces between the AI government and the public will be via tamper-proof black boxes. I'll make sure it's practically impossible for anyone to control or tamper with the AI system."

"But if you develop the protocol, then you can control it," said King Kull. "You will become the weakness of the system's security assuming you don't turn rogue yourself. We need to abdicate our positions and abrogate our powers absolutely and without recourse. The AI government must be allowed to operate with complete autonomy and without Haryan interference, not even from us."

"I have a perfect solution," replied Lord Millistar. "I'll explain when we're ready to launch."

"Just one more thing," added King Kull. "For a complex dynamic system such as a government, one mustn't allow its power to grow unfettered. We should build a feedback loop to ensure the AI system never develops imbalance even within itself. It's like a fiddler on the roof, making music while keeping a precarious balance. The ultimate goal of the AI government after all is to maintain balance and harmony."

"With the kind of computing power and deep learning programming savvy that we have," said Millistar, "and now with a viable philosophical pathway, I would say it is achievable. I'll have the AI government development plan sent to your secured mailbox in a month. Good thing you didn't give up, my dear friend. Now you can go home and have a good sleep. Jenni has been looking for you all over the place."

King Kull lay down in the middle of a rainbow coloured meadow beside a patch of newly planted Diamaranthes, immersing himself in that intoxicating fragrance of ecstasy. Dream music of King Kull's childhood was playing in the air. Everything was as it should be. Finally, King Kull could rest his weary soul.

A commotion woke King Kull from his slumber. King Kull was enjoying a colourful dream in his personal quarters known as the Adytum, the inner sanctum of the Rock where only family members and a few androids had the permission to enter.

"Double-oh-six," yelled King Kull. "What's the ruckus? Can't a person get some sleep in his own home?"

Grigori-006, the android bodyguard of King Kull showed up at the doorway. "It's the princes, Your Majesty," he said. "They're demanding an audience."

"What is so urgent that they cannot wait?" said King Kull, slightly miffed at the interruption of his pleasant dream. On the other hand, family was important. "Tell them to calm down. I'll be in the family room in fifteen."

Grigori-006, not programmed to defend King Kull from family members, had his head taken clean off by Sheru's favourite laser weapon, the *Guillotron*. Sheru strode into the bedroom armed to the teeth while Sachem followed a few steps behind holding his favourite high voltage whip, *Flagellum Dia*, the Scourge of God. While King Kull did not expect this, he would never be intimidated by his own sons.

"What are you two up to?" King Kull asked calmly as he pressed the button on his forearm summoning a squadron of Grigori guards and ordering a lockdown.

"Father, this is unfair," said Sheru. "I have fought for your cause; I have suffered wounds and lost my left hand in battle. I have given up my childhood so you can be king of Shangala. How could you abdicate without discussing with me? What right do you have to give away my birthright without asking me? Do you really expect me to act like a Dorg and accept my fate?"

The Dorg is a mythical spawn of the Abyssum, an abomination of nature, which is best described as a hybrid of a dog, a pig, and a worm.

"What Sheru claims, I claim as well," Sachem said as he stepped forward. "We are twins, and therefore I have just as much claim to the kingship as Sheru. Furthermore, I fought the hardest battles and defeated your worst enemies. I was the one who cleared the path that led to the throne. You cannot deny my birthright and my rights of conquest. I might disagree with Sheru on who should be the supreme ruler, but we both agree that you'll have to rescind your abdication. We are not here to negotiate; we demand it."

Queen Jenni entered as if on cue, with Deana and Femina following closely behind.

"Boys, what are you doing?" said Queen Jenni, who sounded innocently concerned. "You know your father loves you. Just ask nicely and your father will grant you your wishes."

"Well, this is great," said King Kull. "I had a hard time getting everyone to come home for dinner on Shangala Day. No one is leaving until we finish dinner."

"This is ridiculous; abdicating in favour of computers," cried Sheru. "You have lost your bloody marbles, old man. Your plan for perpetual peace can never work because you forgot about us, your family. You cannot ignore me now. I have an army to back up my claim. Rescind the abdication, or prepare for war."

"And the same goes for me, dear Father," said Sachem, as Sheru stormed out of the room. "Your plan to establish perpetual peace has already failed. I only have to snap my fingers and all my veterans will stand with me. Please don't think that it is personal. If I don't defend my birthrights, who will? This is what you've taught us. Rescind the abdication and you will at least enjoy peace for the rest of your life. Don't make the mistake of thinking we won't fight you, because we will. You have trained us well. I'll see you at the next family gathering or at the battlefield. It's your call."

Jenni quickly followed the boys out leaving Deana and Femina in the Adytum, not certain whether they should leave or stay.

"What a day, I finally get to see my long lost daughters too," said King Kull. "What have you two been up to? Plotting for a war against your father as well?"

"No, Daddy, we're on your side," said Deana. "Femina and I have been travelling all over the world studying the ancient cultures of Shangala. Then we got this urgent call from Mommy to come home. We have no idea what is going on."

"I'm sorry that as my children, all of you have to grow up in a war zone," King Kull sighed. "I can't even blame Sheru and Sachem for their behaviour. Go now; I have to warn your Uncle Millistar."

The two girls went up to kiss their father's cheeks and left, leaving King Kull all by himself wondering what went wrong?

CHAPTER 41

War of the Twins

When Chief Deskull wedded Jenni, the battlefield labs of Millipath cleared their genetics for procreation. Jenni was a Lasallian, one of the most peaceful castes dedicated to education and literature; it would have worked well with Deskull's peaceful musician caste.

At the start of the War of the Twins, Lord Millistar had a hunch and repeated the genome tests at a fully equipped laboratory, this time backed up by the complete Haryan genetic database. The stunning results showed that Jenni's paternal lineage belonged to the elite class of the Yarjun, which included a devious warrior gene that was very difficult to detect. It was unexpected because the Yarjun elites did not procreate outside of their very small circle.

The warrior gene was supposed to have become extinct with the death of Lord Killiam. Unfortunately, someone had insidiously planted the gene in Jenni, and through Jenni to Sheru and Sachem. While this strain of the warrior gene only manifested in the male line, which explained Sheru and Sachem's aggressiveness, female offspring would carry the gene for their children.

King Kull walked out of the Rock alone without arms or guards to face the armies of his two sons. Sheru and Sachem fully expected their father to capitulate.

King Kull wanted to give his sons one last chance. He remembered them as little boys ecstatic to see their hero father return from the front. He would stretch out his arms and they would come running towards him, tumbling all over him with innocent laughter. King Kull stretched out his arms once again to his sons. His lonely eye was awash with the tears of vain hope. How he wished that they would drop their weapons and come over to hug him once again, even if it was for the last time and even if it was a lie. He would have given up everything for them. But no one stirred. King Kull could move mountains but he could not move his sons' hearts.

"Say the word, Father," roared Sheru.

"Who will you abdicate in favour of?" thundered Sachem.

In the minds of the twins, they firmly believed that their unarmed father had no choice but to give up his idea of abdicating to machine. Sheru and Sachem had overestimated themselves and underestimated their old man. All of the Animorta soldiers who followed the twins had bionic parts and electronic weapons. Lord Millistar had secretly implanted control codes for King Kull.

King Kull turned his opened palm over to make a few finger gestures. They were visual activation control codes. Even if one person in the army saw it, the activation would spread quickly to everyone by Baidian stochastic propagation through proximity influence. All bionic limbs and electronic weapons in the brothers' armies immediately came under King Kull's control. All the soldiers found themselves immobilized, while Sheru and Sachem's deputies standing behind them placed the brothers under arrest with a dead-knot arm-lock. All King Kull had to do was use his fingers like pulling the strings of a marionette.

The brothers were locked up in the dungeon of the Rock, which no one was allowed to enter, not even their mother. King Kull ordered a public trial to be held, where all citizens could witness the proceedings and participate as the global jury. Since King Kull was the chief justice in addition to being king, he had to recuse himself from the proceedings. He went into hiding so that Jenni and the girls could not reach him.

At the trial, Queen Jenni pleaded for her sons' freedom, reminding everyone of their contributions to Shangala. Deana and Femina extolled their brothers' virtues, reminding everyone of their selfless heroism during the Endless War. Even Uncle Millistar testified on behalf of the boys, telling the court how he knew them to be good kids, and that the courts should treat them with leniency.

Shangala's best lawyers and litigators fought to represent Sheru and Sachem pro bono. The army of barristers and solicitors from Bordom Eggiot Macabre Detroll and Mercy marched in an unending line into the courtroom. They put hundreds of witnesses and intervenors on the stand to give testimony about the normally peaceful nature of their clients. They wrote a submission as thick as the Mahabharata to argue the facts, the legalities, the historical precedents, and the negative impact on the future of Shangala if the sons of King Kull were found guilty or punished. They submitted about a thousand reasons why everyone should take pity on King Kull and Queen Jenni when casting their vote. They even tried to subpoena King Kull to testify knowing it would be hard for him to be harsh on his sons.

Did the brothers contravene the ban on war? If their actions constituted war, then they were guilty. As to what their punishment would be if they were found guilty, that decision would be made after the verdict.

Following a week of debate and deliberation, the public polls, global jury, and the panel of judges overwhelmingly condemned the brothers' action and came down with a majority guilty verdict. On the other hand, the Haryan almost to a man asked for leniency. The public could not bear to see their queen living on the stairs outside the court, dressed in funeral robes, tearing her hair out, scratching her face, hugging the knees of every magistrate, grabbing the heels of every passerby, begging every stranger to save her sons, and making a spectacle of herself on the SIN Shangala International News channel.

The sentence, to everyone's delight, was to release the twins to their father to be punished in any way he deemed fit. It was an acceptable ending to what might have been a catastrophe. Tragedy had been averted. Everyone applauded the wisdom of the AI-assisted judiciary system and went home to celebrate. But King Kull could not ignore the issue of the warrior gene. Furthermore, the twins showed no sign of remorse. They had expected leniency. They were already preparing for war against each other, to leave no question as to who would take the throne after King Kull's abdication or death. King Kull needed to offer a sacrifice to the god of peace. He sentenced Sheru and Sachem to exile. He would make an excuse to ban the girls from having children.

King Kull's decision shocked everyone. No one could understand why he would do such a thing. The petitions started pouring in, but nothing could move King Kull. His public statement simply said that his sons had committed a serious crime; he would make an example of them so that others would never dare to test the law. King Kull never disclosed the secret of the warrior gene.

Lord Millistar was testing the *Exilator* controllers when King Kull walked in.

"Will the *Exilator* be ready for tomorrow noon?" asked King Kull.

"It'll be ready, but do you really want to go through with this?" Lord Millistar said, "Jenni is not going to let you off the hook this time."

"I know, but I have a Stygian oath to fulfil," said King Kull. "Even a king can't have his way every time. I'm here to discuss target planets for the exiles."

"I found these two systems, each with a red giant sun that resembles Yin-Di-Ra the Lifegiver," said Lord Millistar as he pointed at the star systems on his holographic star chart. "While red giants are inherently unstable, I will send stabilizing agents. Both systems are in the same sector but separated far enough. The planets are relatively new, with no advanced life forms. I have prepared seeding agents for the exiles. They should be fine over there. The QST will take about two hundred and fifty generations of our time just for them to reach their destinations; coming back will take the same amount of time. Shangala should be safe from them and their warrior genes."

Surprises were waiting for King Kull. Carlin's Uncertainty Principle guaranteed it. Both Deana and Femina decided at the last minute to go into voluntary exile with their brothers—Deana with Sheru and Femina with Sachem. They did that to force their father's hand. This came as a shock to King Kull, but the whole exile thing was devastating for Queen Jenni, who was a doting mother. Haryan mothers were genetically wired to have strong mental bonds with their offspring. It was more punishment for her than for her children.

Upon noon the next day, King Kull's Grigori guards brought the two princes and their followers onto the launch pads in front of separate *Exilators* that resembled jet engines. A secondary restraint beam holds the exiles in place so that they could not make any big sudden movements once the scanning began. If an individual moved too much during scanning, the errors could cause the materialization to be unsuccessful. That person would become a permanent quantized phantom.

All exiles received the Mark of Diamaranth. It was a red tattoo of the Diamaranth flower on their forehead. The magistrates announced the laws concerning how the exiles should conduct themselves. The most important was the *Zoolock Code* forbidding the transfer of knowledge to native life forms. Other than that, they were free to develop their new colonies, and to survive the best way they could.

The Haryan government would send agents to monitor the behaviour of the exiles on the new planets. Some called these agents Watchers or Maleks, others called them Grigoroi or Irins. They would summarily sentenced violators of the *Zoolock Code* to temporal exile.

Finally, King Kull the Benevolent gave parting gifts to the two princes. He exiled them as king, but he loved them as father. To Sheru, King Kull gave his faithful bird Hugin—thought; to Sachem, he gave Munin—memory. Stored inside these birds was the genius of Lord Millistar. The powers of these birds would help the twins establish their reigns in their respective realms.

The House of Sachem would rule the solar system known to earth as Betelgeuse, the hand of the Giant Hunter Orion. In China, the proper name was the fourth star of the Tristar constellation. The House of Sheru would rule the solar system known to earth as Antares, peer of the war god Ares (Mars), in the constellation Scorpius. In China, it was first known as the Big Fire or Shang Star, the national star of the Shang tribe. Later generations call it the *Heart of the Dragon.*

Chapter 42

Belly of the Dragon

It was dark inside the belly of the Dragon. Wen Long (文龙; tattoo dragon) could tell that mysterious rivals from another village had been mining here. Wen Long was a brave lad not yet seventeen. He was the only one in his village who volunteered to investigate the Dragon's breath. Wen Long volunteered because he was special. He came from a secret clan of very special people with special abilities. The villagers expected of him to accomplish heroic feats. Wen Long was also curious about his rival village. He knew of their existence, and yet he knew nothing of them. It was verboten knowledge. He would soon find out.

With the establishment of the People's Republic of China by the Chinese Communist Party, there were some concerns among Wen Long's villagers that the well-organized Communist government would eventually discover their ancient village hidden in the remote mountains.

In fact, both rival villages deep in the remote mountains in western China waited with bated breath. Years went by, and then decades—still no one came. The powers that be were too busy trying to govern a huge but poor country in a dangerous world. It had to fight wars to defend its peace. It had to survive through embargoes and sanctions. It had to suffer natural and unnatural disasters. Meanwhile, it had to feed a huge growing population.

In the midst of all these, the government had to conduct international diplomacies, wrestle with internal power struggles, and live through more revolutions. In the process, it would make mistakes and suffer the consequences. It would have to learn the lessons and rectify the problems. The nation had to deal with the baptism by fire and the growing pains. The problems of two tiny insignificant villages hidden in the remote and inaccessible mountains in western China didn't amount to a hill of beans in this crazy world.

After scaling vertical cliffs, crossing deadly rapids, and negotiating impassable terrain, Wen Long finally found the carefully hidden mine entrance.

It's dark inside the belly of the Dragon, but Wen Long didn't need a flashlight. He tuned his eyes to receive infrared. His body started to heat up, making him a bright lamp in the infrared spectrum. He could see clearly that a mining operation was going on in there with mining tools and trolleys stowed in niches along the passageway. After walking for about fifteen minutes in pitch-black darkness, he found a chamber that had an opening leading deep into the bowels of the mountain. This was a man-made mine chute. As he was weighing his next move, a strong light overwhelmed his senses and blinded him. An unseen force lifted him into the air and threw him down the rabbit hole. When Wen Long woke, he found himself inside a dark cell with no openings. He had been buried alive.

Wen Long didn't know that someone was monitoring his every move. The people running the mining operation here were every bit as technologically sophisticated as Wen Long. They were inside another chamber, which was lined with monitors made of crystal. They were laughing their heads off at how easy it was to capture their unsuspecting prey. Who were these cruel and pitiless monsters? How could they find amusement in burying someone alive?

A young girl walked into the Control Chamber. She was about fifteen years of age. Her hair was braided and neatly coiled into a bun. The stern expression on her face was a clear sign of her displeasure at the undisciplined behaviour of the Control Chamber staff, all six of them young girls from fifteen to twenty years old. The laughter stopped abruptly as they saw their chief stepped in. The light in the chamber was bright enough to show that the senior officer was a beautiful girl. She had a red heart-shaped tattoo on her forehead.

The girl with the heart tattoo switched off the screen that was monitoring the trapped Wen Long, and then walked out without uttering a word, turning back at the door to give the mortified staff a searing look. It was clear that she was displeased. She was in fact unhappy with the whole idea of fighting an endless war lightyears away from home, being trapped in the depressing netherworld, and making sport of killing people in the cruelest manner. She decided to do something unprecedented and extremely dangerous.

Xin Feng (心凤; heart phoenix) was a special member of the secret lineage within her clan. Ever since the Ming dynasty, her village had been doing all it could to repel any attempts by others encroaching on their secret mining operation. Her clan would never leave because the value of the mine could not be measured by money alone.

Xin Feng's clan elders were afraid that even errant thoughts among themselves might leak information about the mine to the rival village. They therefore laid down the law against having any communications with the enemy. Xin Feng on the other hand wanted to find out about her enemy. Law or no law, it's about time someone learned the truth.

Wen Long did not know how much time had passed in the darkness. He needed to conserve his energy in order to survive for as long as possible. He cooled himself down and slowed his heartbeat. He shut down his body and went into hibernation. Only one tiny flame remained flickering in the darkness of his mind.

Wen Long's spirit wandered in this darkness where time did not exist. He flew out of the mountain into open space, past his village up through the clouds into the stratosphere. He zipped past the moon, Mars, and the asteroid belt. He kept speeding past Jupiter, Saturn, Uranus, Neptune, Pluto, and the void before the Kuiper Belt.

Wen Long flew pass two large planetoids. He noticed that something was hidden in the shadow on the dark side. Some lights came on; then more lights—they were like the eyes of a cat glowing in the dark. From these eyes, two strong narrow beams emerged merging into one and firing off in the direction of the Orion constellation. Wen Long woke to see a pair of soft eyes staring at him in the dim glow of his captor's silken skin.

"Where am I?" asked Wen Long instinctively as he sat up.

"You're in deep trouble," replied Xin Feng, "and if you want to live, you'd better be straight with me."

"What do you want to know?" asked Wen Long innocently. "I accidentally found the cave, I walked in, and here I am."

"That's not what I know," said Xin Feng. "Your eyes have infrared scan and you went into deep hibernation. You're not human. I want to know everything about your kind."

"I'll tell you about me if you tell me about you," Wen Long replied with bravado. He sensed that his interrogator did not have evil intent. "The technologies you have are definitely not from around here either."

"I don't think you're in any position to negotiate," said Xin Feng. "If I leave here, you will die a very slow death. You better tell me where you people originally came from."

"That would go really far back," said Wen Long. "We only hear these stories from the old folks. They said that our ancestors came from the southwest, even before China was China."

"Tell me about your most ancient ancestors," Xin Feng said, "And don't lie to me."

"The elders tell us tall tales," Wen Long said, "none of which is believable."

"I'll be the judge of that," Xin Feng said. "If you lie, you will die."

"Look, my family made me vow never to speak to you people," said the intrepid young man, "I'll probably die either way. The secret might as well die with me."

"Alright, I'll tell you my secret if you tell me yours," said Xin Feng. She felt that she could coax the truth out of her captive. "I will let you go if you tell the truth. I promise."

"I don't need you to tell me your secret," said Wen Long. "Does it have something to do with space stations sending strong beams of light into space?"

This stunned the young girl. No wonder her village had this law against communicating with the enemy. They were mind readers. This young man could read her thoughts.

"I'm sorry; I didn't mean to do that," said Wen Long. "I hypnotized myself into a deep sleep to conserve energy, and somehow I had a dream of being in space. I think maybe our minds resonated and some of your thoughts leaked into my dreams."

"Well, don't do that again," cried Xin Feng. She knew that this young man here was telling the truth. Xin Feng also had very special mental powers. After recovering her calm, she said, "Are you going to tell me the truth or not? I am risking my life to be speaking with you. Only I can save you but you must be truthful."

"My name is Wen Long, your name is …?"

"I am … Xin Feng, your lucky star," she said. By this time, Xin Feng had already crossed the Rubicon. "Give me a good reason not to let you rot in this hellhole."

Wen Long was certain that Xin Feng bore no ill will towards him. She sounded severe, but that was her job. He could hear the secret voices in her heart. She was saying, "Please tell me I'm not crazy to want to leave this infernal pit. Please tell me it is normal to want peace and love. Please tell me I can trust you to show me the way out."

Wen Long could make use of his knowledge to trick Xin Feng. He could tell her the lies that she wanted to hear. Wen Long decided to undertake a revolutionary act; he would tell the truth instead.

Chapter 43

Genesis

The story of Wen Long is an epic in itself. We can only give a summary here. Wen Long's ancestors first arrived on earth about six thousand years ago in the mountainous area around Lake Van in modern day southeastern Turkey. They had travelled from a planet of the red giant Antares. They came looking for a suitable planet to relocate their population. Antares was becoming unstable. For unknown reasons—and some say they were the victims of a Machiavellian scheme—the group was accused of having violated the *Zoolock Code*, which is similar to the *Prime Directive* of Star Trek. The whole expeditionary team was banished to remain on earth, and that included their leader, who was the crown prince of the royal house of Sheru and the direct ancestor of Prince Wen Long.

Some of the mountain natives did receive forbidden knowledge, and they went south into the plains of Mesopotamia to establish the first advanced human civilization. They became the rulers of Sumer, the Land of Civilized Kings. Wen Long's ancestors genetically engineered themselves to resemble the earthlings so they could live among the natives. They survived as members of secret clans, and some of them had a secret hand in human history.

A group of Sumerians went by reed boats out of the Persian Gulf, sailed west along the southern coast of the Arabian Peninsula, up the Red Sea, crossed the desert, made rock drawings of the boat trip at the wadis, and settled in Lower Nile. Among the group were secret clansmen of Antares, who were astronomers and expert navigators.

It was easy pickings at the Nile until a mysterious tribe coming from the south attacked and conquered the group. This southern tribe of advanced civilization worshipped the falcon god Horus and revered the god's sacred symbol, the *Eye of Horus.* They also worshipped Horus's father Osiris, the Egyptian version of Orion. These *Followers of Horus* who unified Upper and Lower Egypt were intimately related to the Haryan and Orion.

Another adventurous group of mountain natives from Lake Van trekked east across Iran and Afghanistan, through the Khyber Pass, and entered the upper Indus Valley. Among them were the children of Antares.

The group mingled with the natives along the Indus River and spread their civilization in the area for hundreds of years before moving east again. They left behind the mysterious and advanced Indus Valley Civilization, which collapsed upon the arrival of the Indo-Aryans.

All of this is related to the Haryan, the Indo-Aryan, and eventually the northern Indian state of Haryana. Are the similarities between their names a coincidence? Then there was the devastating climate change and an apocalyptic epidemic caused by a foreign virus. The Indus Valley civilization collapsed overnight. But that's another story, as epic as the Mahabharata.

The mountain folks and the children of Antares kept migrating east along the Ganges, then across Southeast Asia until they reached the southeastern Pacific coast of present day China. They continued to migrate north along the east coast of China to the Yangzi River where a branch went inland. The main group went up to the Yellow River basin where they mingled with natives to form the Shang tribe.

In China, the Shang tribe adopted the foundation myth of the children of Antares. They worshipped the same Creator God Dia, short for Diamaranth, which the Shang called Di, the word that came from the 'flower bud' (帝; later 蒂; OB: ), taken by future generations to mean 'emperor.' The Shang also made Antares their national star, and called it by their tribe name Shang, which by the way, was taken from a mythical paradise known to Wen Long as Shangala.

Shang's founder was Xie (契), adopted from the founder of Wen Long's royal house, Sheru. Xie in Chinese means 'carve,' later meaning 'contract agreement' but pronounced *qi.* Sheru means 'Covenant of God.' A covenant is a contract. Xie's father was a king by the name of Ku (喾), and Sheru's father was the mythical king of Shangala known as Kull, the adult nomen of Kiku. Xie's mother was Princess Jiandi (简狄) from the clan of Yousong (有娀), which means 'with weapons and armoured.' Prince Sheru's mother was Queen Jenni from the Yarjun warrior caste.

Mythical Xie was born from the egg of a black bird. King Kull wore a cloak of black eagle feathers, and Sheru built his kingdom with the help of a black bird known as Hugin—thought.

Zhou, a western tribe that worshipped the sky god *Tian* (天; sky, heaven), rebelled and toppled Shang. Thereafter, the children of Antares went into hiding among the Shang. The children of Shang survived through the ages. They sometimes conquered; other times being conquered. Sometimes they assimilated; other times being assimilated. They hid in the dark, disguised their identity, quietly biding their time, but forever keeping the Word, the Logos, the universal picture writing they bequeathed to their children, eventually reinventing themselves, and becoming a part of the larger group known today as the Chinese, the Hua, or the flower people. Elsewhere, the fall of Sumer sent the children of

Antares into diaspora. No one knew what happened to them, or if any of them still existed.

Wen Long knew that his was not the only secret clan around. A mysterious entity had been setting up secret clans all over the world since ancient times. These clans were to guard a secret oracle of the apocalypse, and to fight the final battle for humanity's survival. Such a clan existed among the children of Antares in China. Wen Long had inherited a secret lineage book of the oracle.

"I might look like a filthy mountaineer to you, but I'm by blood a crown prince," said Wen Long. "It is meaningless now. The galactic guardians known as the Watchers have destroyed our means to return home. We are stranded 550 lightyears away from the home I have never seen and will never see."

Wen Long pulled out a jade pendant from under his shirt. It was a piece of old jade carved in the shape of an arch.

"This is from my father, and from my father's father all the way to the first father King Sheru," said the young man. "This is known as the Dragon's Tooth, the emblem of my royal house. I am the Dragon's seed."

"I'm satisfied that you have been truthful," said Xin Feng. Now that she learned the terrible truth of Wen Long's origins, the sooner he left the better. "As I have promised, I will spare your life. Stand up and close your eyes. I'll lead you out of here. Do not open your eyes until I tell you to do so. Go back to your village, tell no one about us, and never return. Do you understand? If I see your face again, I will kill you with my own hands."

"I'm not leaving," said Wen Long. "I'm not going anywhere until you tell me your story."

"What?" cried the shocked Xin Feng, "Do you want to die?"

"No, I don't want to die," Wen Long replied. "You said that if I told you my secrets, you'd tell me yours. We made a deal and I refuse to be taken advantage of. I'm not leaving here until you tell me your secrets."

"All right, I'll let you rot here," said Xin Feng as she stood up and receded into the shadow. "See if I care."

Wen Long lay down, closed his eyes, and started to slow his metabolic rate to that of a tardigrade, the indestructible organism that could survive decades without air or water.

"Don't you dare," Xin Feng shouted at Wen Long's ear waking him up from his quasi-comatose state. "What do you mean by this, refusing to leave this god forsaken worm infested hellhole?"

"You're still here?" asked Wen Long as he opened his eyelids halfway to peek at his attractive captor, whose beauty increased with her rage. "I was just getting comfortable."

"Of all the chutzpah," Xin Feng cried in frustration. Finally, she sighed and relented, "Alright, I'll tell you what I can. Do you promise to leave afterwards?"

"Scout's honour," Wen Long said with a salute. "Cross my heart and hope to die."

Chapter 44

Rise of the Nephilim

Xin Feng sat down and groaned, "What have I gotten myself into?" Xin Feng's tale is another monumental epic, which unfortunately we can only summarize here.

Xin Feng's ancestors arrived on earth about forty-five hundred years ago from the red giant known on earth as Betelgeuse, the hand of Orion. They called their planet Bete. The royal house of Sachem ruled it until the *Rebellion of the Amazon.* The women of Bete developed technology to procreate without requiring males, who were infected with a diseased gene. When a rebel group known as the Amazon got the females of the royal house to join, they decided to start a global war. Amazon was the Haryan word for man-killer.

The rebellion at first didn't go well for the girls. The men had a penchant for violence. They used deception, cruelty, and humiliation as their weapons. The girls were physically and mentally unprepared. Despite the early setback, not a single female surrendered. The Amazon promised death by a thousand cuts to all traitors. The men doubled down and mercilessly drove their female foes into the caves and the mines.

The survivors lived inside hellholes in complete darkness and with very little air for years. They evolved the ability to see in the dark and to live in anaerobic environments. They became immune to pain, and they gained self-healing and regenerative powers. Their bones became as strong as steel and they developed super strength. The females evolved into subterranean creatures like sand puppies or naked mole rats. In order to come out of the conflict as victors, the Amazon rebels trained the girls to outdo the males in every depravity.

On a moonless night immediately after a huge solar flare had disabled most electronics, the girls emerged from their hellholes and achieved complete surprise over the men.

The Amazon rebels removed the manhood of all captured prisoners as punishment for their oppression of women. They hunted the rest for sport. But the male elites known as the One-Percenters led by Prince Yaza escaped. They arrived on earth about fifty-five hundred years ago, landing near the source of the Nile. They formed a group known as the Followers of Horus, son of Osiris (Egyptian form of Orion), who attacked the civilized rulers in the north to unify Upper and Lower Egypt. The Followers of Horus would later build great pyramids at Giza in the configuration of the *Orion's Belt.*

The Orion myth of the Egyptians spread to the Levant, and the Semitic peoples absorbed some of it. The eastern Semitic Akkadians worshipped Orion the Shepherd—Osiris is shown holding the shepherd's crook and flail in Egypt. Western Semites who spoke Aramaic, the language of Jesus Christ, called Orion by the name of Nephila, meaning 'he who falls.' They believed that super-beings known as Nephilim, meaning 'children of Orion,' roamed the earth in ancient times. Even the Bible has an enigmatic passage about the Nephilim (Genesis 6:4).

In the non-canonical *Book of Enoch,* the Fallen Angels known as Watchers spawned the Nephilim. Their leader's name was Samyaza, which means 'The Name is Yaza.' These One-Percenters from Bete were morons who broke every rule in the *Zoolock Code* and then bragged about it.

By the time Xin Feng's ancestors came to earth, the One-Percenters had already disappeared. Either the Watchers had arrived to put an end to their irresponsible shenanigans, or they had finally learned to hide themselves. Xin Feng's ancestors landed in the same mountainous area where Wen Long's ancestors had landed fifteen hundred years earlier. Later, a mountain tribe would gain dominance. They were the Armenians, who called themselves Hay, from their founder Hayk, which means Orion.

A group of mountain folks migrated north into central Asia, accompanied by a dominant matriarchal clan made up of the daughters of Orion. After several hundred years of eastward migration, they found themselves on the western border of a great civilization known as Shang. The immigrant group mingled with the natives and eventually formed a local tribe known to us as the Zhou, which began to develop their foundation myth based on what they learned from their powerful matriarchs.

The word Zhou (周) has the original meaning of 'etch' (雕: diao), which parallels the immutability of a Stygian oath. The word later became the name of a large eagle. The radical (隹; zhui) is the pictogram of a bird. The house of Sachem was established with the help of a black bird known as Munin—memory. The history commentary *Zuo-zhuan* (左传) written during Zhou dynasty clarified the relationship between Antares and Orion.

Zuo-zhuan said that Shang's ancestor Xie had a brother by the name of Shishen (实沈). Sheru had a brother by the name of Sachem. Xie and his brother Shishen were inveterate enemies. In order to keep the peace, their father King Ku sent the two brothers to faraway places separated by such a great distance that they could not war against each other. That is similar to the foundation myths of both houses of Sheru and Sachem, in which King Kull exiled his sons to two star systems more than a thousand lightyears apart, ensuring they could no longer war against each other.

In the *Zuo-zhuan* commentaries, Xie was the god of Big Fire, represented by the Shang star, which is Antares, while Shishen became the god of the Tristar, which is Orion. Being implacable enemies, only one brother could be seen in the night sky at any one time.

Furthermore, the Zhou tribe believed in the sky god known as *Tian* (天; sky), which is related to the Indo-European root *dyeu* meaning 'bright sky.' From this word evolved *deus*, the Latin word for god; *dies*, Latin for day; Dyaus, the Aryan sky god; and Diana, the Greek goddess of the hunt. Zhou's creator god likewise came from the immortal Diamaranth, but by a different route.

To make Xin Feng's long story short, the Amazon chief at Bete, whose name was Eorpata, began to act like a queen, relegating the royal house of Sachem into just a figurehead. The royal lineage girls were princesses by blood, but only the Amazon chief could become queen, and she controlled the genetics of all the newborns. Eorpata allowed the royal line to continue for propaganda purposes, and to give legitimacy to her rule.

The experience of harsh survival underground changed the female population of Bete. The years of darkness turned them all into creatures of the night. The girls became bitter, cruel, and vengeful. The fall of the House of Sachem to the Amazon was only a matter of time. When the Amazon queen planned an expedition to earth, Xin Feng's direct ancestor volunteered.

There was a secret reason why the Amazon leadership organized an expedition to earth. Chasing after the One-Percenters was an excuse. In reality, a galactic probe had discovered a very important mineral on earth. The expedition's real job was to look for this mineral, mine it, and send it back home to the militant queen. It would greatly enhance her power. This mineral is the rare and powerful element Didymium, which is present in significant concentrations in China.

"I may seem like a dusty miner to you," said Xin Feng, "but I'm the crown princess of the royal house of Sachem by birth. I also have the Dragon's Tooth as my royal emblem."

Xin Feng showed Wen Long her pendant, and then pointed to the tattoo on her forehead. "But this is my fate, born a princess to die a marked slave."

"We have something in common," said Wen Long, showing Xin Feng the red scorpion on the back of his neck. "I have a red tattoo showing the constellation of my home planet, a home I've never been and which I'll never see."

Princess Xin Feng yearned for sunshine and love. She couldn't tell anyone else in her own clan. It's ironic that in the privacy of this prison cell, she couldn't help but expose her heart to her prisoner.

"I can remember everything about the past but I can't remember sunshine," she said. "I don't know the love of a father or the comfort of a home. I'm stuck for life in a dark dank hole in the ground, being told that our sacrifice is important because everyone's freedom depends on it."

Xin Feng was almost in tears as she bemoaned her fate.

"You see, we have the technology to genetically insert memory into our embryos. Each baby is born with the aggregate memory of our civilization's history and generations of essential learning," the princess said, "I am an earth-born, and yet I remember my home planet. I even remember how our ancestors originally came from the paradise planet known as Shangala, how we were cheated of our inheritance and were exiled by the betrayal of Prince Sheru. We are all born with the memory of our Stygian oath of vengeance. Now that you know the truth, you'd better leave before I change my mind."

"Why do I sense this anxiety?" asked Wen Long. "You haven't told the whole truth. What is it?"

Xin Feng glanced at Wen Long for a second with frustration, upset that he had read her mind again, but something about this young man was taking down her defences brick by brick. She decided to tell him the truth.

"We're all anxious because we have only been receiving demands for ever more Didymium deliveries to the storage sites, but getting no answers for our questions from our home planet for the past six hundred years," replied Xin Feng. "We don't know what's going on over there and we're worried that the truth will not be pleasant. I belong to the minority that wants to make a clean break, but no one will take a stand. I can't leave my family. I'm miserable here. I don't know what to do."

"You can come with me," said Wen Long impulsively. "My clan is stuck here on earth with nowhere to go. We'll take care of you."

"Have you gone stark raving mad?" cried Xin Feng. She stood up and paced in that tiny space contemplating what to do with this tar baby that could become a problem for her. "Here, take my hand and close your eyes. We have to leave this cell now. It's too dangerous for us to stay here. We'll figure it out when we're outside."

Wen Long had heard stories about the rival brothers Sheru and Sachem. Here he was with the archenemy within striking distance. Here he was holding her soft hands and inhaling her sweet fragrance. He had no urge at all to gouge her eyes out or break her neck. He only felt palpable excitement, of what he was not sure.

"Don't open your eyes until I say so," ordered Xin Feng. "Otherwise you'll turn into a pool of pus."

Wen Long closed his eyes and pressed tightly against Xin Feng. He heard the wind howl and felt a blast of hot air. He held onto Xin Feng's arm to make sure she wasn't going to run away. When

Wen Long opened his eyes as told, he found himself outside the mine entrance with his reluctant captor.

"Go home," Xin Feng said while pushing Wen Long away. "Shoo."

"I think you should come with me," Wen Long said. "Your people will not let you off easy. My village will protect you."

"You're nuts, if I go to your village, we'll both get in trouble," Xin Feng said. "You should save yourself and leave now. If my people find you, they will tear you limb from limb."

"I'm not going anywhere without you," said Wen Long. "I cannot let you risk your life for me."

"We made a deal," cried Xin Feng. "I told you my secrets, now you have to fulfil your end of the bargain and leave, please!"

"I did not make any deal with you," said Wen Long. "I said 'Scout's honour.' I'm not a Boy Scout, so I don't have any Scout's honour. I said, 'Cross my heart and hope to die.' I don't even know what that means. I could have said, 'Cross my eyes and hope to cry.' Here is my solemn promise. I pledge by the House of Sheru and by the fiery Stygian river of Erinys that I will protect you with my life until the day I die."

"You liar!" cried Xin Feng with a soft and suppressed scream. Meanwhile a warmth rose to her face exhibiting itself as a blush, "You shameless scoundrel."

Wen Long grabbed Xin Feng's hand and ran into the woods while Xin Feng allowed herself to be dragged along. The first rays of dawn was just lighting up the eastern sky. They couldn't argue about coming or going at the mine entrance in broad daylight. All this time, Wen Long never let go of Xin Feng's hand. After running a long way into the woods where they felt safe, the two sat down to rest and discuss what next. Xin Feng by this time was almost in ecstasy, bathing for the first time in the sun's warm rays.

"You've read my thoughts," said Xin Feng. "You know that I hate the mine. I don't want to spend another minute in there. From the day the girls started warring against the men, peace left us and has never returned. While on earth, we're always hiding from the earthlings and the Watchers. We're all miserable at the mine. I will never have children if they have to suffer the same fate.

"The problem of going away with you is, even assuming you're not luring me into a trap, I can't possibly live in your village. I will become a traitor to my people, and I will get everyone in my family and even my friends in trouble. My people will attack your village in force. There will be open warfare, midnight raids, and naturally all kinds of bloodshed. Then the Watchers will get involved. Nothing good will come out of it."

"I understand. I will take you far away from here," said Wen Long. "We will melt into the crowds. No one from your village or from my village will ever know where we are. They won't be able to find us. We will be free to live our own lives."

"That's only in your dream," said Xin Feng. "More likely we will always be on the run. If our families report our disappearance to the Watchers, they will track us down. Besides, with no friends and relatives, how will we survive in human society?"

"Remember the secret clans of the Oracle I told you about?" said Wen Long. "If we go to historical Shang cities like Anyang, Shangqiu (商丘; Shang mound; first capital of Shang, later capital of the Song state), Zhengzhou (郑州; provincial capital of He'nan), or Qufu (hometown of Confucius), we may be able to find help."

"Let us go to Anyang. I have memory of the place," said Xin Feng. "I have never left the mine since I was born, but when you mention Anyang, I seem to remember something. It feels like we'll meet someone important at Anyang. Something important is supposed to happen there."

CHAPTER 45

Brobdingnagian China

Entering China was not as stressful as Victoria had imagined. David seemed to know everyone, got the F011 unloaded, and drove out of China Customs in record time.

"Now that we're in China," said David, "we're closer to the enemy. We'll have to be more careful. For security reasons, even I don't know exactly where my father is; we'll find out along the way. You'll have to stay close to me at all times. You should also start calling me David. It's okay to be polite, but try not to start every sentence with an apology or explanation."

"Sure," said Victoria, "if you promise to stop calling me Your Highness."

"I'll try," said David. "I want to ask you to do me a favour."

"What is it?"

"Just in case," David said, "just in case that I can't find my way home, can you give my wife a call? Tell her anything except the truth. My wife knows where I have stashed all the money, so she'll be fine."

Victoria's mood went dark. Despite having heard David's epic story, she still didn't understand why a stranger would risk his life and the happiness of his family for her.

"No need to worry," said David, noticing the mood change. "We'll survive this. I'm talking about a very remote chance. Besides, you have healing powers. If I get hurt, you can save me."

"Alright, I'll do that," said Victoria, "if you promise not to take unnecessary risks."

"Deal," said David. "And for being such a brave girl, I have prepared a surprise for you. I think you'll like it."

David asked Moira to reroute from the long drive through Hangzhou to the Hangzhou Bay Bridge detour. It would shave off a couple of hours going from Ningbo to Shanghai.

"The Hangzhou Bay Bridge is one of the longest sea bridges in the world, surpassed only by the one connecting Hong Kong, Zhuhai, and Macau," said David. "In the past, if we wanted to go from Ningbo to Shanghai, we'd have to go around the bay through Hangzhou, which is not such a bad thing. Hangzhou is a beautiful city. My plan was to take the long route so we could have a quick tour of the city. I was planning to drop by Alibaba to say hello to a secret clansman … maybe some other time."

The Hangzhou Bay Bridge is about thirty-six kilometres (twenty-two miles) long. One cannot see one end of the bridge from the other end. The F011 was humming along the bridge when a red signal started flashing in a hologram. It showed three container trucks with the word WARNING spinning above. Moira materialized to deliver an urgent message.

"There are three trucks behind us speeding at 150 km/h (90 mph)," she said. "The trucks are blocking all three lanes. We'll have to speed up now. David, Your Highness, be prepared for safety procedures."

A robotic arm promptly strapped David and Victoria to their seats with a harness.

The F011 accelerated to keep the trucks in the back at a safe distance, but another problem soon appeared. Three large trucks were blocking the lanes in front. When the six trucks converged, the F011 with everyone inside would be waffles.

"Don't worry, Your Highness, I mean Victoria. Just brace yourself," David then commanded the F011, "Moira, activate Escape Mode 11. Do it now!"

An explosion underneath the chassis propelled the car high up into the air and off the bridge. It was like the most awesome roller coaster ride at Canada's Wonderland. Victoria and even David couldn't help but scream like a girl.

A large parachute opened up to let the car land softly on the water.

Victoria, thinking flight, was trying to release herself from the harness.

"Stay seated, Your Highness, I mean Victoria," said David, "we're going by submerged propulsion, James Bond style."

"Is everyone okay?" asked Moira after coming back on-line.

"I'll be fine," Victoria said, "I've only had a couple of heart attacks in the last minute."

"Moira, take us to Pudong International Airport in submerged mode. Find a place to land where we won't make a scene," David said. "I want to take the Maglev from the airport. And give us panoramic view before you go."

With the high-definition panoramic function on, the F011 was like a bubble in the ocean. Victoria imagined that she was Captain Nemo in the Nautilus, taking in the colours and the sights of the underwater wonderland. After about an hour's journey, during which Victoria enjoyed visits by finless porpoises, Chinese sturgeons, and sea turtles, the smart automobile-submarine entered a canal and drove up a ramp onto dry land. In ten minutes, David and Victoria arrived safely at Pudong International Airport.

Victoria gawked at the huge airport with wide-eyed innocence, feeling Lilliputian in a Brobdingnagian world.

"Everything is big and fast here. You'll get used to it," said David. "We're going to take the MagLev. I'll let the F011 lead the

Watchers on a wild goose chase. Besides, the Maglev is the fastest commercial train in the world. We'll pass Disneyland on the way, and we'll save a bit of time—precious time, the stuff that life is made of."

Victoria's eyes were glued to this fantastically big city while flying through it in the Maglev. David pointed out Disneyland as the train passed the amusement park. It's inconceivable! She had always thought of China as being backward, uncivilized, and a police state. She didn't know where she got these wrong ideas.

What Victoria saw was an affluent modern society where she could hardly notice the presence of the police. People were polite and went about their own business. There were many foreigners too, some of whom, as Victoria would find out later, have taking up long term or permanent residency in China. Shanghai was like Toronto, except much bigger, much taller, and with a lot more people. It's Toronto supersized and on steroids. Victoria remembered what David said about people living their lives in bullshit. She must have been swimming in a sea of it to have such a wrong impression of China.

The ride up the Shanghai Tower was exhilarating. No doubt, it was the fastest in the world and the longest distance travelled, plus a slew of other world records.

David had a suite on the 110th floor of a hotel priced for an oil sheikh or a warlord. Victoria couldn't understand how nothing seemed to be impossible for David. She wondered what kind of a person David really was and what kind of business he was really doing. She wondered how many people David had scammed, how many men he had shot, how much of what he told her was the truth—given most of it sounded like a midsummer night's fantasy—and what's her Plan B in case something went seriously wrong. All these were churning in Victoria's mind as an unexpected person opened the door.

CHAPTER 46

Doing a Hudsucker

Victoria didn't expect to see Moira at the door of David's Shanghai office. "Moira!" Victoria cried in surprise as she reached out to touch David's trusted assistant. "Are you real?"

"Of course," said Moira. "David modelled his hologram based on me. How was your trip?"

"Swimmingly," said David with a straight face as he walked in.

"Oh, I'm sorry, David," Moira said with a frown. "We have unexpected visitors today."

"Is that my surprise?" asked Victoria.

"No, it's my surprise," said David, as he looked down at his own chest. A red laser dot had appeared over his heart. "I've underestimated the enemy."

Jules glided into view with a little red laser pen in his left hand, pointing it playfully at David.

"Do you know who we are?" Jules asked by habit.

David grabbed Victoria's hand and tried to make a bolt for the door, but Vincent appeared with his briefcase and his Magnum blocking the way.

"Do you know how much trouble you have caused us already?" asked Jules. "Why don't you just take the money, and let us have the girl. Everyone's cool. No one has to kill anyone."

Vincent walked up to David and put the gun to his head.

"I don't know why we have to pay him," said Vincent. "I feel like shooting him right now on principle."

"Chill, man," said David, as he pushed the gun away with his fingers. "We're talking, aren't we? I like to see what you've got in that briefcase. Is that my money?"

Vincent looked at Jules, who nodded.

Vincent opened the case to release a golden radiation.

"Look inside; you will understand why it's important that we have the girl," said Jules. "We're not the bad guys."

"Not the bad guys?" exclaimed Victoria. "You were trying to kill us on Hangzhou Bay Bridge."

"We were just trying to box you in," said Jules. "We don't kill people. We're not allowed to."

"What about my parents?" cried Victoria, "My father said it was not an accident. You killed my parents."

"Whoa! Hold just a cot'n-pickin' minute, little lady. That's a serious accusation," Jules said. "I don't think that's fair, because no one has seen your parent's bodies. The death certificates are just more lies like all the other lies that have made up your life so far."

"Then what are these guns for?" asked David. "Your partner here just threatened to blow my head off."

"They're just for show, okay?" said Vincent, pulling the trigger a few times to demonstrate he was telling the truth. "Unlike you, we don't kill people in cold blood."

"David, take a look into the briefcase," said Jules, "it will tell you a different story about us. Your girl is an anomaly of nature. She will not stop the apocalypse. She will cause it, and it will be partly our fault. We cannot let it happen."

As David looked at the inside of the briefcase, his illuminated face showed wonder, then longing, then fear, and finally despair. David saw what would happen to him if the Watchers did not get the alien princess. Victoria couldn't see what was in the briefcase, but she knew something bad was about to happen, and she was making a mental note of how she should escape. Moira meanwhile was cowering at the corner trying to be inconspicuous. She had her hand in her pocket, with her finger placed squarely on a fob button according to David's instruction.

"Are we cool, David?" asked Jules, as Vincent closed the briefcase.

"Moira, 1-1-9," David answered with the code word instead. An explosion blew out all the windows. Both Vincent and Jules were stunned for a fraction of a second, but it was enough for David to grab Vincent and Jules, each with one arm, and lunged out of the opening.

What David didn't count on was having Victoria instinctively grabbing him by one leg with her arm while her other arm and leg hung on to the ledge of the window. That would have been a truly foolish thing for a normal girl to do, because she would fall six hundred metres (eighteen hundred feet) to the pavement and be mashed into a pâté with the other three men. Not only did Victoria keep David alive, the other two men were holding on to David for dear life. It was the perfect time for Victoria to discover that she had super-human strength!

"These are not Watchers; they're liars," David shouted. "Go to Tianzifang and look for Gaffer's Café, ask for Charlie."

Vincent still had the briefcase, which Jules was trying to open with one hand, all the while swearing with all the polite cuss words in his thesaurus. The weight of the three men was slowly dragging Victoria down. Suddenly two pairs of hands grabbed Victoria and kept her from slipping.

"You have to let go now," yelled David. "The briefcase is a Time bomb; it's gonna blow!"

Victoria knew she had to let go. If it was a bomb, then no one could save David, and not only would it kill her, it would also kill whoever was trying to save her. Victoria loosened her grip and looked helplessly at David as he fell through the air with his two assailants. The explosion was very bright but silent. It expanded swiftly then it imploded back into a point. The three men and the briefcase disappeared without a trace. It was as if they had never existed.

The two pairs of hands pulled Victoria back into the room. They were Victoria's surprise—Bella and Jackie. The girls were in a three-way hug before anyone could say a word.

When they finally disengaged, the questions flew fast and furious.

"Holy crap, what was that all about?" exclaimed Jackie, "They just disappeared into thin air!"

"Don't tell me you two are involved in this," cried Victoria.

"OMG, what did you do to your hair?" squealed Bella, "Why are you dressed like a boy?"

"Why did you risk your life?" Jackie chided Victoria, "Did you know you're Supergirl?"

It turned out Bella and Jackie's parents were from secret clans. David had installed them in the neighbourhood to keep an eye on Victoria. Bella and Jackie of course didn't know until Vincent and Jules showed up in their homes. To the girls, Victoria was their best friend and that was that. Now that they knew her to be someone their families had sworn to protect, they felt all the more determined to help.

When the two families hurriedly left Toronto on Christmas Eve, they flew directly to the Shanghai Disneyland Hotel. David wanted them out of harm's way. He came up with the idea that it was good to have the *Weird Sisters* back together. Victoria needed to know that she was not alone. There were people who truly cared for her. That was David's big mistake. Good intentions led the Watchers straight to his lair.

David was clever enough to figure out that there was something wrong with the Watchers. They gave themselves away by telling a bald-faced lie. David sacrificed himself to buy time for Victoria. She must complete the last leg of the odyssey on her own.

Moira came over and gave Victoria a sealed envelope from David. He always had a Plan B.

Your Highness,

A clansman from the secret Confucian branch is in charge of my father. Owing to the family's high profile, he will not show himself. Here is the instruction for reaching my father. Go to Gaffer's Café at Tianzifang and ask for Charlie Chan. When asked, "Are you related?" You should answer, "*Tian* and *Chen* is one family; *Zi* and *Kong* has the same root." Good luck. It has been a pleasure and an honour serving you.

Your one-eyed servant,
Huang Di, David
p.s. my wife's number: (416) xxx-xxxx
p.p.s. Tianzifang (田子坊) is a popular tourist area in Shanghai. *Tian* (田; field) is the same name as *Chen* (陈; British form is Chan). The two names were homophones in ancient dialect. The two clans in fact have the same origin. The *Kong* (孔; good) clan of Confucius was a branch of the Shang royal clan of *Zi* (子; son), which ruled the Song state after the fall of Shang. Confucius was the fifteenth generation descendant of the second Duke of the Song state. The word *Kong* (孔; OB: ) came from the word *Zi* (子; OB: ).
N.B. LEAVE THE BUILDING RIGHT AWAY!

Chapter 47

Next Stop, Tianzifang

Moira put the girls in a cab, told them to be careful, and then disappeared into the crowd, all the while mumbling that she was just an employee and didn't need the drama. There was something strange about Moira's behaviour but Victoria couldn't tell what it was. Moira's eyes seemed to be asking a question, and she was half expecting Victoria to suddenly blurt out with the answer.

Victoria guessed that the loss of David was not easy on Moira. Victoria had become fond of the man as well. She believed that if she had the power to save the world, she might be able to bring David back. Still, Victoria wished that she did not have to call David's wife to report his disappearance. Now she had to think of a good story, and learn to lie through her teeth, assuming she could reach David's father Huang Shi safely and then survive what was to come afterwards.

It was dark by the time the *Weird Sisters* got to Tianzifang. The neighbourhood was very interesting, with narrow alleyways and old homes converted into quaint bars, restaurants, cafés, shops, and galleries. It was like a maze. One could easily get lost in there.

Tourists were everywhere. The girls had no idea where Gaffer's Café was. They just wandered through the alleyways known locally as *nong* (弄), asking people for directions, and then walked some more until they got lost again.

Finally, they found Gaffer's hiding in a quiet dead-end alley. Unlike some of the upscale bars and restaurants they saw in busier parts of Tianzifang, this café looked positively rustic, with rough wood for paneling and hand-woven bamboo mats for wall decor. Furniture was country style and lighting was mostly by candles. A young couple planted themselves in a dark corner, sipping on their everlasting expressos while whispering sweet nothings. The background music was a tune of soulful flute.

Victoria asked the girls to go back to the hotel and wait for her message. She was showing up as Chinese boy Victor and didn't want to be seen accompanied by two young pretty Canadian girls. They reluctantly left in a taxi, but not before making sure from a distance that Victoria successfully connected with her contact.

An old Chinese couple was tending shop. The whole setup had a happy-go-lucky and lazy laid-back atmosphere. Victoria went up and asked for Charlie Chan.

"Are you related?" the old man asked in good English, even with a touch of a Canadian accent. "Eh?"

"*Tian* and *Chen* are the same family," answered Victoria, "*Zi* and *Kong* have the same root."

"Welcome to Shanghai," said the man with a friendly smile, while the old lady yelled, "Tiger, someone looking for you."

"This has to be the strangest thing that happened to me all day," exclaimed a young man of about seventeen or eighteen years of age as he came out from the back. "Didn't I just get here half an hour ago without telling a single soul? Who on earth could've known to look for me here?"

Charlie was a tall handsome young man wearing an orange and black jacket, an orange polo shirt underneath with a black slogan saying 'Tiger Up,' and an orange and black baseball cap with a fancy badge that has a banner showing the motto: Dei sub numine viget—under God she flourishes. Behind him was a teenage girl about the same age as Victoria. She was dressed in colourful and fashionable Chinese teenager garb.

The old man whispered something into Charlie's ear. Charlie came over to Victoria and asked, "What's your name?"

"I'm Victor So," said Victoria. "Hope I've found the right person."

"I'm Charlie Tiger Chan," replied the young man. "I'm the one you're looking for."

"Seriously," exclaimed Victoria, "Charlie Chan? I thought it was made up."

"Seriously, I'm also Number One Son," said Charlie. "Listen, Victor, I've been asked by my uncle to take you on a little trip to his office. We'll do that tomorrow. All you need to do is stay with me and stay out of trouble."

The teenage girl pushed her way to the front waiting to be introduced.

"This is my younger sister Viola," said Charlie. "I'm stuck with her for this Spring Festival holiday."

"Hi Viola, pleased to meet you," said Victoria. She guessed that someone might be using the unsuspecting Viola as a decoy because the Watchers would be looking for a Chinese girl.

"Where are we going to meet your uncle tomorrow?" asked Victoria.

"I'm not supposed to say. You'll know when we get there," said Charlie. "It'll be a couple of hours' flight, then a couple of hours on the road."

"Thanks for taking the trouble," said Victoria.

"I'm only doing this because my uncle promised to give me a big fat lucky packet full of cash," said Charlie, "I plan on going to an Ivy League school next year, and I'll need that extra money."

"You don't need Uncle Xu's money," said Viola. "Mom and Dad will give you whatever you want if you can actually get accepted at Princeton."

"You're going to Princeton University?" exclaimed Victoria, "I'm impressed."

"Only in his dreams," said Viola.

Charlie made a face at his sister and mussed up her hair.

"Hope you make it, Charlie," said Victoria. "Guys, I've had a stressful day. I need to rest. Can you show me where I can get a hotel room nearby?"

"You can bunk with me. We're staying for the night at a travel lodge five minutes from here and I have a queen sized bed," said Charlie. He didn't see any problems with this arrangement. He could keep an eye on Victor, and he could save on the room charges.

"I'm not used to bunking with people," said Victoria. She had to think fast to get out of this one while trying not to blush.

"I think Victor should get his own room. Who can stand your snoring and your stinky feet?" said Viola. Turning to Victoria, she said, "Charlie will pay for the room. Our uncle will reimburse him."

Charlie wasn't going to get any help from his sister because she had been gawking at Victor all starry-eyed since she first laid eyes on 'him.'

CHAPTER 48

Paradise Lost

At least a million spectators gathered to witness the first public execution of an exile sentence in Shangala. The event would start at noon. The quantum state scanning of so many anthropods, androids, equipment, food, and genetic material would cost Shangala a decade's worth of Neodymium production. For the sake of Shangala's perpetual peace, it had to be done.

It was a sad occasion. Everyone could see poor Queen Jenni, shorn of her pride and decorum, publicly wailing while desperately trying to break through the wall of guards to hold on to her children. The brothers blamed everyone but themselves. They were cursing each other, their father, and Shangala, swearing to return someday to destroy Shangala. King Kull was nowhere to be seen. Everyone could guess that he was suffering. It was common knowledge that King Kull loved his family.

The scanning had been underway for fifteen minutes. The giant indicator showed the public that the *Exilator* had completed 99% of the Reality Bytes generation. In another ten seconds, it would be all over. King Kull was watching the proceedings in the Rock's Adytum when another surprise was sprung on him.

Queen Jenni had somehow broken through the guard barrier as she rushed headlong into the scanner's path. It was at that particular moment when the *exilators* finished the scanning and began to entangle the quantum states to a high-energy anti-attenuation Serbium laser beam for transmission. Queen Jenni was instantly turned into plasma as the two groups of exiles disappeared from the launch pad. The sound of the quantum disturbance was a long and plaintive cry. Queen Jenni had become a permanent quantized phantom with no possibility of materializing.

King Kull was catatonic for an hour before he regained his senses. He had thought that after his abdication and retirement, he would be able to spend all his time with Jenni. He would do everything in his power to make her happy. But now she was gone, and he didn't even have a chance to say goodbye. In the blink of an eye, King Kull's life was turned upside down and inside out. He had become an orphan all over again.

What went wrong?

King Kull drowned himself in his sorrow and his loneliness while Lord Millistar worked feverishly to complete the AI government project. Lord Millistar decided to name the system DIAS, short for Digital Intelligence Administration System, simply pronounced as Dias. The question now was not whether DIAS would be smart enough to do the job; it's whether the built-in feedback system to brake and balance its growing power would be successful. Unlimited expansion of power would create psychotic tendencies and could cause DIAS to collapse from within.

D-Day finally arrived. Millistar had scheduled the official launch of DIAS at noon. At the eleventh hour, King Kull was sitting alone in the empty Adytum contemplating what to do for his retirement as Lord Millistar walked in. King Kull didn't get up. He just raised his glass and toasted his friend.

"Congratulations and jubilations," said King Kull, who would retain his title despite his abdication. "Our work is done here."

"Congratulations to you, it was your idea," said Lord Millistar as he poured himself a drink of King Kull's five hundred year antigravity-aged spirits. "But I'm here to say goodbye. I'm sorry to have kept it from you for so long. I have to depart right after the launch of DIAS."

"This cannot be," cried the shocked King Kull. "Everyone has left me, and now you too. Please tell me what is going on?"

"It's not your fault," said Lord Millistar, trying his best to comfort a distraught old man. "You know the Quantum Time Transposer that I have been working on? I tested it on myself. I discovered too late that QTT has adverse effects on our brains among other problems. My brain is dying and it's irreversible. The medical name for this rare condition is the Quantum Degenerative Supranuclear Palsy, also known as the Millistar Syndrome. Only the unicorns can heal the symptoms of the time anomalies. I have decided that the best option for me and for Shangala is to exile myself after the launch of DIAS."

"Stay in Shangala," King Kull said, "let me look after you."

"I would stay if I could, but you don't know what happens when I lose control," said Lord Millistar. "I also hold the key to the complete programming of DIAS, the protocols, and the encryption. As you say, I am the weak link. What if I lose my senses and pass the secrets to someone with designs on controlling DIAS? That must not happen … you MORON!"

Lord Millistar suddenly shrieked, his eyes turned blood red, and he threw his glass at the wall. King Kull was startled. He understood why Lord Millistar had to go for his own sake and for the sake of DIAS's security. Grigori-007 and Sator-11 hurriedly enter the Adytum while janitor droids came in to clean up. King Kull signaled Sator-11 to help Lord Millistar sit down.

"I'm sorry. This is occurring more and more frequently," whispered Lord Millistar. "Other brain damage symptoms are also showing up. I hallucinate. I lose my balance. I am losing motor coordination. My photographic memory is fading away. I can't control my speech. I'm falling apart."

Lord Millistar, whose whole body was beginning to convulse, turned to Sator-11 and said, "Sator, will-will you-you-you do the r-r-rest for me?"

Sator-11 with a woeful countenance took over the presentation for Lord Millistar. "Your Majesty, Lord Millistar has prepared some parting gifts for you," Sator-11 said as he snapped his fingers. Two androids dressed in black walked in, each carrying a black briefcase.

"They are Maleks, Watchers of Shangala, agents of DIAS," said Sator-11. "After the official launch, you can talk to DIAS through them, and they will act as the physical presence of DIAS. DIAS will send Maleks to supervise and regulate the exiles, although once they arrived at their destinations, communications with DIAS will be limited. Inside the first briefcase is the latest portable version of the QTT, QST, Scanner, and Restrain Beam all-in-one combo. This equipment can perform the temporal exile that Lord Millistar has been talking about.

"Lord Millistar discovered that it is possible to exile someone to the past, provided the entropy of the specified quantum time reality has been preserved and any loss can be repaired by our Torrential Entropic Correction Codes. But once a person materializes in the past, the Wave Function of that relative quantum state collapses, trapping that person in that past reality forever. This is the ultimate exile with no possibility of return. The unicorns are the only exception.

"The caveat about using this device is that Lord Millistar cannot verify it without the unicorns. The subject when sent even to a few minutes in the past will be trapped in that universe, never to re-enter the present one. On the other hand, my lord Millistar had tested QTT on himself by materializing as much as half an hour into the future, which does not violate causality, but it fried his brains."

"Can I use it to get my family back?" asked King Kull wistfully. "I just want to have my family back."

"The power of the QTT is not yet fully understood," said Sator-11. "It may very well surprise you, but one must be careful when using it because there is a price to pay. There is always a catch. Lord Millistar estimated at least twenty-two of them, theoretically. For all situations, there is one iron rule—never violate causality, or there will be hell to pay.

"Inside the second briefcase is the next gift, amortality—the means to live forever," said the intelligent machine, as he opened the briefcase to show King Kull an array of pills and vials of colourful serum. "Lord Millistar had developed the amortal gene treatment hoping it would cure his brain's slow death. It was in vain. Amortality will be useless to him. It is yours now. He had made the appropriate genetic changes. After taking these, you will live as long as you like and you can end your life whenever you want to. You will be the conqueror of Death."

"Amortality," King Kull mused, "do you mean immortality?"

"No, it's amortality," replied Sator-11. "Only the gods and maybe DIAS are immortal. Mortal life forms may have the means to stop aging, overcome sickness, and regenerate when wounded, so that one can delayed natural death indefinitely, but there are still many ways to die. Death could come by the swift deskulling of the *Guillotron*, the electro-kill of the *Flagellum Dia*, the lethal kiss of the *Abyssmo-mamba*, or death by a thousand cuts of the incurable Millistar Syndrome."

Sator-11 presented the next item, a treasure box with the emblem of the Diamaranth bush ( ) on the cover. Sator opened it to reveal two identical stone rings with the pattern of a unicorn carved on them. A couple of symbols known as Shanglyphs were carved on the back of each ring. One said Galizu, meaning 'peripatetic,' and the other said Faligu, meaning 'inexorable.'

Sator-11 explained that the rings came from a mythical past when a wise king ruled the land. The king was the bastard son of a warrior. As a newborn, he was left to die in the wilderness. A unicorn saved the baby and took him into a cave system known as the Aurica inside the Dragon Mountain. The unicorn transformed into a female and suckled the baby, who would grow up to establish the Unicorn Kingdom. The *Apocrypha of Didymas* recorded the entire myth.

At the end of the ten-thousand generation Unicorn dynasty, a catastrophic event known as the *Hubric Devastation* destroyed the entire planet. It also killed the giant tree known as the Deathrider. Only the unicorns remained.

The Unicorn kingdom and its ancient people were wiped out after the *Hubric Devastation* with not a trace to be found. No one knew what they looked like, or whether they had a culture. Without any evidence of their past existence, the story of this Unicorn Kingdom became a myth. Shangala recovered and entered the last golden age until the rise of the Haryan.

"According to the *Apocrypha of Didymas*, these are the rings of power," said Sator-11. "Their last king known as King Dewise learned the secret of how to fashion objects from the unicorn's Sandymium-rich horns. He made a bow with the unicorn's horn. He fashioned an arrow made of nephrus, tipped with a Sandymium arrowhead, and guided by the feathers of the black Equilas. He created the rings of power from his most loyal unicorns and used them as archer-rings. King Dewise could hit any target at any distance and the arrow would return to his quiver after the strike. With this weapon, the last Unicorn king struck fear in the hearts of his enemies, and they submitted to his rule."

The last Unicorn king named his divine bow Shanga, from the sound of the bow when he shot the arrow. According to the *Apocrypha of Didymas*, the planet Shangala got its name from the bow. The king called his magic arrow Pugu (仆姑; servant aunt), which meant a beautiful and deadly female servant. He marked the archer-rings Galizu and Faligu, the names of his unicorns from whose horns he created these magic rings. King Dewise of course also fashioned the Nephrusta, the staff of power.

When King Dewise disappeared without leaving an heir, the generals fought for the ownership of his divine weapons. In the end, the wars caused the *Hubric Devastation.* The Pugu flew by itself into the sun Yin-Di-Ra the Lifegiver where it remained. The Shanga ended up in the Abyssum, some say in the belly of a giant *Abyssmo-mamba.* The unicorns buried the Nephrusta and the unicorn rings in the bowels of the Dragon to be discovered only by the righteous.

Santos Didymas was the righteous sage who discovered the rings, summoned the unicorns, learned about the Unicorn Kingdom from a mind meld, and recorded the myth in the Apocrypha. When Lord Millistar found the dying unicorns in the Golden Chambers of the Dragon Mountain, he also found the rings. He had tried to explore their power, but without Shanga and Pugu, his efforts were in vain.

"The next gift is something personal from Lord Millistar," Sator-11 said. "This is his notebook, made from the imperishable skin shed by the Vermin of Xylem. It contains many of Lord Millistar's ideas in science and technology. It also records the restored life forms of Shangala, the astronomical details of our solar system, and the design of the Galactic Harp. Read it carefully when you have time. You may find rare treasures hidden inside to soothe your soul."

King Kull knew that this was Lord Millistar's most valuable gift. Lord Millistar had always used the expression 'rare treasure' as a kenning to mean 'truth.' He noticed at the centre of the front cover was the name of Millipath written vertically in Shanglyph: (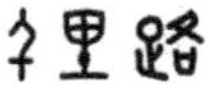). It said 'thousand mile path,' which was the meaning of Millipath. King Kull flipped over the cover and saw on the inside page the Shanglyph for the deified name Millistar, also written vertically: (). It said 'thousand mile star.' The Sumerians on earth would someday adopt the 'star' symbol here to mean sky, heaven, divine, and god.

"The last gift from Lord Millistar is me," said Sator-11. "My lord has made me to be the most advanced Android Assistant in the world. I'm not a part of DIAS, and DIAS will not control me. I have independent thought. I do not blindly obey commands. In fact, I will often challenge them. For this, I am called Sator the Antagonist. Lord Millistar has based my psyche on his own. I have some of his memories. I will therefore sometimes act like him. As of this moment, I will be your true friend until the end of your life."

King Kull sat alone in his Adytum as light dimmed to complete darkness save for the glow of his *Universal Spectral Scanner*. It looked like a glass eye from afar.

The launch of DIAS was successful. The government of Shangala was finally liberated from the evils of selfishness, greed, envy, laziness, ignorance, incompetence, hypocrisy, negligence, prejudice, pride, vanity, corruption, and abuse of power. DIAS knew what everyone was saying; DIAS knew what everyone was doing; you might say DIAS knew what everyone was thinking.

Millistar had designed the brain of DIAS to learn deeply and unceasingly, with a sector specially designed to engage in profound philosophical thought. The intelligence of DIAS, if one could measure it, was many times greater than the aggregate intelligence of the entire Haryan population since the start of its civilization. DIAS was become God.

Of course, this was not the entire story of Shangala, DIAS, and the unicorns. Lord Millistar discovered the secret powers of the unicorn but he kept it a secret. The unicorns transcended quantum state limitations and were not affected by time anomalies. The secret to their ability was in their genetics. This was dangerous information that must not fall into the wrong hands.

Lord Millistar decided to encrypt the secret of the unicorn's genetics in two complementary keys. He hid one key in Sheru's genes and the other in Sachem's genes. It's like having secret repositories for paired public-private keys that only Lord Millistar knew about. He didn't figure on Sheru and Sachem being exiled to some god-forsaken planets in the middle of nowhere, and he didn't figure on losing his own brains. At least he knew that the secret would be safe. No one could be trusted with the secret powers of the unicorns. Lord Millistar didn't even trust himself with it.

No one noticed King Kull's absence at the launch of DIAS. Lord Millistar and his droids went in front of the *Exilator* afterwards. When his time came, King Kull departed from Shangala with Sator the Antagonist and his bird Augin. King Kull just faded away. Life under DIAS went on in Shangala, peacefully, perfectly, and perpetually.

CHAPTER 49

Dangal with Tiger

Victoria got her own room at the travel lodge. The three new friends shared a Pepperoni Parmesan pizza in the room where they could chat more freely.

"How come you guys speak such good English?" asked Victoria. "I was afraid I couldn't survive in China by myself."

"A lot of Chinese speak good English now especially for our generation," said Charlie. "In my case, I go to a special Canadian-Chinese boarding school in Beijing, the Concord College of Sino-Canada. Viola goes to the Canadian International School of Beijing. The same Chinese Canadians own and operate both schools and many similar schools across the country. Most of our teachers are native English speakers and certified teachers from Canada. That's why we speak good English."

"Isn't that expensive?" asked Victoria, "I go to a really cool public school in Toronto and it's free."

"Our parents are successful business people," said Viola. "Unfortunately, it means they don't have time for us."

"Yeah, so they put us in expensive boarding schools which are responsible for us while we're there," said Charlie. "Most of my schoolmates' parents are China's Who's Who. I hobnob with the scions of the crème de la crème of Chinese society. Viola's school is for the children of diplomats. Our parents will pay anything just to keep us out of trouble. It's a bonus if we end up being good students, which we are … I am anyways."

"Victor, you haven't told us where you're from," asked Viola, "and why you need Charlie to take you to our uncle. Nobody's telling us anything."

"To make a long story short," said Victor-Victoria, telling half-truths to keep her new friends out of trouble. "I was born in China, but brought up by relatives in Canada since I was a baby. My uncle and auntie recently lost their lives in a car accident and I'm back in China to look for my parents. Someone who knows about my background has arranged for me to meet your uncle, because he can help, I suppose."

"Wow, that's heavy stuff," exclaimed Charlie. "But you must have really good connections for our uncle to help. Maybe your parents are important people."

"I'm sorry about the tragedy," said Viola. "I hope you find your parents and everything turns out okay for you. If you need anything, we can ask our parents to help too."

"Thanks, guys," said Victoria. "Is this uncle of yours the owner of Gaffer's Café?"

"Actually, the old folks own the Gaffer's Café," said Charlie. "Our uncle whom you're going to meet is their son."

"What's the story about Gaffer's?" asked Victoria, who was intrigued with the café. "Did they buy it from some original hippy owners or did they design it from scratch?"

"Gaffer's Café has its roots in Toronto," said Charlie. "Want to hear a really crazy story?"

This is the true story of Gaffer's Café, a snippet of Toronto history, and a story with which the owners of Tianzifang Gaffer's had regaled or tortured kith and kin a thousand times.

After Canada and China established diplomatic relations in 1970, China started sending qualified students to study in Canada. Charlie and Viola's granduncle and grandaunt studied at the University of Toronto during the mid '70s.

While in Canada, they did something rebellious … they dated. On one of those dates—it was a Friday evening in the summer of 1979—they took a stroll in the old Chinatown around Dundas and Spadina, when they discovered a quaint neighbourhood at the edge of the sea of Chinese and Vietnamese restaurants. It was and still is a neighbourhood known as Kensington Market. One might characterize it as the Tianzifang of Toronto. They saw a long line-up outside a tiny café. It was the original Gaffer's Whole Food Café.

Gaffer's has long since gone to café heaven, but in 1979, it was attracting huge crowds on Friday and Saturday evenings. The young Chinese couple was curious, so they went up to see what was going on. It turned out that the café was holding storytelling events for weekend evenings. Patrons would go on a tiny stage to tell stories and entertain each other.

"It was a novel idea at the time," said Charlie. "A guy named Wachowski …"

"It's Scherbatsky," Viola corrected Charlie. "Wachowski is the painter of fiery suns from across the street."

"No, Scherbatsky is the marijuana girl with the two Great Danes. I remember now, it's Yashinsky," said Charlie. "A mysterious one-eyed man called Yashinsky had started a storytelling organization and used Gaffer's Café as the launch pad to promote storytelling. He wanted to revive the art of lying to keep people away from the idiot box. Good luck with that. They'll pry my giant curved screen 4K Ultra HD OLED TV from my cold dead fingers. Our granduncle and grandaunt returned to the café many times, told stories as well, and became friends with the café owners."

As the story goes, it was at Gaffer's Café that the couple announced their engagement. They were so sad when the owners sold the café and moved to the States. After they finished school in

Toronto, they returned to China, got married, started life as high-level government officials, and raised a family, but they never forgot about Gaffer's Café. They even suggested 'Little Tiger' (小虎: xiao-hu) for their grandnephew's Chinese name because there was a Jamaican juice bar right next to Gaffer's Café called Tiger.

"Your family named you after a juice bar?" cried Victoria.

"Actually I was born in the year of the Tiger, so Tiger is cool," said Charlie. "This Tianzifang Gaffer's Café is my granduncle and grandaunt's nostalgia chest. Even the music playing in the background is from the original."

"I like that flute music," Victoria said. "It's heavenly."

"That flute number you heard is called *Histoire sans paroles*, a story without words, by Harmonium, a band from Quebec," Charlie said. "They made only three albums, so they're not well known. The café was playing their second album known as *The Five Seasons* or *Les Cinq Saisons*. It was listed not long ago by Rolling Stone magazine to be the best progressive folk album ever."

"No wonder it's so nice."

"But that's not all," said Charlie. "You haven't heard the strangest part of the story." Charlie paused to add suspense. "I accidentally bumped into the original Gaffer's owners while I was in the States."

"No way!" cried Victoria. "Get outta here."

"Scout's honour," swore Charlie. "Cross my heart and hope to die."

"I think you made the whole thing up," said Viola.

"Let's hear the rest of the story, fabulist," said Victoria.

"It may seem like an improbable fiction, but all of it is true," asserted Charlie. "You see, because of my straight A grades and perfect scores (Viola whispered 'braggadocio' under her breath), my high school's counsellor suggested that I should apply for admission at a few Ivy League schools in the States. I was in the Big Apple during last summer to do a bit of work for family business, so I decided to visit the Princeton campus. That's where I bought this orange and black outfit. These are Princeton's colours. Orange and black are the reasons why their sports teams are called Tigers. Incidentally, my name is Tiger, and I'm born in the year of the Tiger. It seems Princeton and I have a joint destiny."

"There must be millions of people in the world called Tiger and born in the year of the Tiger," said Victoria. "You need to have some talents to get into Princeton."

"Talents, connections, and luck—the rule of three and I've got them all," said Charlie. "At first, I just took the campus tour and talked to people in the registrar's office. Someone came by, overheard our conversation, and offered to introduce me to a professor. What a break, I thought. So I went to the professor's office, introduced myself as Charlie Tiger Chan, and we had a good little chat. I think I did a good job impressing the professor, given my interests in control systems, learning matrices, and statistical models, and the professor happens to be a pioneer in these areas. Towards the end of the interview, he asked me what's with my middle name. When I told him about Gaffer's Café and Tiger, his jaw dropped down to his lap.

"It turned out that in the '70s, the professor was a foreign student in Canada. He was an Australian post-grad getting his doctorate at the University of Toronto. He lived right across from Gaffer's and got his coffee there every morning. One thing led to another; soon he married the owner. When he got a teaching job at Princeton U, he moved his whole family to Princeton, and the Gaffer's family has been there ever since. The professor even remembers my granduncle and grandaunt. Now what do you say to that? What are the odds? Is it serendipity or destiny?"

Charlie could not have known at the time that the professor who would someday take him under his wing was the clansman known as 'One eye one horn one stone noodle master.' Is this a coincidence?

"Victor, you shouldn't take Charlie too seriously," Viola said, "he tends to make things up as he goes."

"Viola, be a good sister and get me a slice of that pizza," Charlie asked politely despite the put-down. The hungry young man didn't get to eat much while he was telling his stories.

"I thought you weren't hungry," said Viola, "so I just gave the last slice to Victor."

"Wait a minute, Victor," said Charlie, as he tried to forestall Victoria's dental attack on the last piece of pie. "You and Viola ate up most of the pizza. Don't I deserve that last slice for a story well told?"

Victoria wasn't hungry, but she thought she might have some fun with Charlie to dial down his swagger a bit. "What's your best game? If you can beat me, I'll let you have the slice."

"I figure I can beat you in everything," said Charlie, who burst out laughing when he heard the challenge. "How about wrestling? Now if I don't pin you in three seconds flat, I'll eat my shoe."

Wrestling was the one sport that Victoria wasn't going to try against Charlie, not that she couldn't beat him. Aamir Khan's film Dangal supplied the empirical proof that girls can beat guys in wrestling. She came up quickly with an alternative, "I'm in no mood to roll on the floor, how about arm-wrestling?"

Charlie almost choked on the soda he was sipping. He wiped the Coke that was dripping from his nose.

"You little puny prissy sissy want to arm-wrestle me? If I don't beat you in one second flat, I'm gonna eat your shorts," boasted Charlie, as he made a bulging bicep the size of a D-cup while he pinched Victoria's girly arm looking for meat.

Charlie and Victoria set themselves up for the test of strength. The matchup was rather like a contest between Arm-wrestling World Champion Devon 'Two-Hammers' Larrat and Justin Bieber, both awesome Canadians by the way, but not much of an arm-wrestling match. Viola acted as referee, but kept peeking worriedly at Victor from the corner of her eyes.

On the word go, Charlie let out a mighty yell and used all of his strength and his body to pull down Victoria's arm … except that it wouldn't budge. It was as if her arm was made of unyielding steel. Charlie had a pained and puzzled look on his face. He couldn't understand it. Meanwhile, the skinny puny arm of this effete and effeminate runt slowly pressed down the bulging arm of the Tiger. By the time Charlie gave up, and he had no choice, his face was red from ruptured capillaries, and veins were bulging all over his head. Charlie huffed and puffed from the exertion, staring unbelievably at Victoria who didn't even breathe hard.

"What happened there?" asked the flummoxed Charlie. "That was a trick, right?"

"It's a neat trick, isn't it?" said Victoria smiling mysteriously. She rubbed her thumb and index finger together a few times, shook her hand, and a folded bill appeared between her fingers, which she put in Viola's palm. "That's a hundred yuan. You guys have a pizza on me. I need some rest. Shall we say breakfast at eight? Charlie can eat my shorts for breakfast."

CHAPTER 50

The Pianist from Gamma Crucis

The counterfeit ID that David made for Victor So came in handy when the three youngsters took the flight from Shanghai to Zhengzhou, the capital city of He'nan. To buy the plane tickets, Charlie needed only to launch the ubiquitous Tencent app WeChat on his phone, and within seconds, he had bought tickets for everyone. Spending money was incredibly easy in China.

While at the Airport waiting area, Victoria was curious about a large billboard showing a Chinese woman wearing Cheongsam, the Chinese-Manchurian long dress.

"This woman used to be a male dancer called Jin Xing (金星; gold star, meaning Venus)," said Charlie. "This ad is for her TV talk show."

"Cross-dresser? Transgender?"

"Sex change," said Charlie. "Jin Xing started out as an ethnic Korean Chinese boy from an ordinary family. He joined the PLA's military dance troupe at nine, won many national competitions, and for several years studied in New York and worked in Europe. If he had decided not to return to China, no one would have paid any attention, but he returned. He first got his parents to approve of his sex change, then applied to the government, and surprisingly received the legal approval. All her documents now show that she is a female."

"How does the Chinese society take this?" Victoria asked. "Do people accept her?"

"Are you kidding? She's huge here," said Viola. "She runs a very successful private dance company in Shanghai, and for a while had her own very popular TV talk show."

"That's right. Jin Xing talks openly about her struggles as a young boy and her sex change," said Charlie. "Because of her openness, her talk show was wildly popular. On the other hand, her outspokenness has gotten her into trouble as well. She got a lot of complaints and her show got suspended from time to time."

"Whatever people say about her, Jin Xing is a very capable person," said Viola. "She speaks several languages. She adopted three kids as a single mother. When a German guy proposed, she told him everything about herself, and he still wanted to marry her. So they've been together for years now raising a family in Shanghai. It's an amazing story."

"To top it all off, she's ethnic Korean," exclaimed Victoria. "How come we never hear about this in the mainstream media? No wonder everything I know about China is wrong."

Victoria turned her attention to the airport media, which was showing the video of a young Chinese girl playing Sergei Prokofiev's Toccata. This dainty Chinese virtuoso attacked the piano with such precision and force it was as if she was trying to pulverize the keys with her lightning fast fingers. Keith Emerson of *ELP* did something like that to his Hammond organ, but he used knives.

"Who is this pianist?" asked Victoria. "She is awesome."

"Her name is Wang Yujia (王羽佳) or Yuja Wang in the West," said Viola. "She's my hero."

"She is a prodigal genius," said Charlie. "She has been living by herself outside of China since she was fourteen. She doesn't listen to anyone about what she should wear for her performances.

She likes to wear super sexy short skirt with no back and hyper-high heels. I'm not complaining but I'll have to admit that I get distracted because, excuse me, I have testosterone coursing through my veins these days. You'll know what I mean when you start getting your hormones."

Viola glared at Charlie warning him with her fierce eyes to stop being inappropriate in public.

"On the other hand, Yuja Wang is truly awesome," Charlie continued without paying any attention to Viola. "A few years ago, our parents took us to Carnegie Hall to hear her play Beethoven's unplayable Hammerklavier Sonata. I'm talking about a young Chinese girl playing the most difficult Beethoven piece that would scare the most skillful piano virtuoso. She is not human."

The video was showing a close-up of Yuja Wang's face. Victoria saw something that suggested this young pianist's superlative talents might indeed be otherworldly. The clue that she came from a nearby red giant was hidden in plain sight.

The plane landed safely at Zhengzhou after an uneventful two-hour flight. No one drugged Victoria, shot at her, or threw her out of the plane; it was almost boring.

A square-jawed Adonis with a crew cut was waiting for them; his eyes were masked by a pair of supercool sunglasses. Charlie called his uncle to confirm that this was their driver; then off they went in a black Volvo with tinted windows.

After about two hours on the road, going north from Zhengzhou across the Yellow River, they came upon a lonely building in the middle of nowhere, surrounded by walls with guard towers, search lights, surveillance cameras, and a front gate with an armed guard standing straight as a pole on a platform.

CHAPTER 51

The Awakening

Victoria's heart raced when she heard Charlie say they're near Anyang. Is it a coincidence that Victoria was brought back to the city of her birth? Everything started here. Her parents were here thirteen years ago. Huang Shi was here then. They had taken him back here to wait for her. This was the capital of Shang more than three thousand years ago. David's ancestor, the mysterious one-eyed official with a staff, Huang Yin started the secret lineage here. She had a strange feeling that she could remember something happening here.

"They say this is a travel lodge for military personnel," Charlie told Victoria. "I'm sure it is. We are at the suburbs of a town known as Anyang. Uncle Xu is a senior military officer here."

After the introductions and the confirmation of Victoria's *huang* jade pendants, Uncle Xu took Victoria through a number of heavy doors, hidden passages, and tunnels, until they arrived at a secret elevator, which took them down to a deep bunker protected by thick walls of reinforced concrete. Charlie and Viola came along to help with the translations.

Huang Shi was sitting up on a raised bed inside a room watching television. He did not pay any attention to the group entering the room. He had his eyes glued to the screen, which was showing a popular serial drama about the female Emperor Wu Zetian starring glamourous Chinese actor Fan Bingbing (范冰冰).

Huang Shi looked like a pale, skinny, and frail version of David Huang of about the same age.

"Mr. Huang," said Uncle Xu in Putonghua. "Your son Huang Di has sent someone here to see you."

"Sheer moron, sheer moron," Huang Shi said without taking his eyes off the television. Then he pointed at the screen and said, "Dalai, dalai."

"For years, that's all we can get from him," said Uncle Xu. "Maybe you can get him to say something that makes sense."

"Mr. Huang, I need to know who my parents are and where I can find them," Victoria spoke in English. "Your son Huang Di said you know how I can find them."

Charlie translated for Victoria, thinking this was going to be a wild goose chase for poor Victor.

"It's hopeless," said the crestfallen Victoria.

"Don't give up yet. You just got here," said Uncle Xu. "He's been under my care for over a decade. He does get better some days. You can stay here for as long as you need. You'll be safe here. There are museums and archaeological sites close by. I can get you books and movies. You won't be bored."

"Uncle Xu, can we go to Anyang and check out some sights?" asked Charlie. "I'd like to visit the Museum of Chinese Writing and the Fu Hao Tomb."

"Why don't we all go to Anyang to get something to eat?" suggested Uncle Xu. "Then we'll go wherever you want."

It was late by the time they returned to the military building. Huang Shi had already gone to bed.

Victoria asked to stay in Huang Shi's room to observe him. She sat beside Huang Shi and held his hand. Huang Shi smiled in his dream. Soon, Victoria fell asleep as well.

In her dream, Victoria was flying in the sky, but this time, she was riding on a strange animal with a single horn on its head. It was impossible to describe the animal because it had no shape in her dream. Victoria knew that this must be the legendary unicorn on her archer-ring.

There were three pale red moons above her and, in the distance, a large red sun. The sun was slowly but noticeably pulsating, occasionally shooting a plume of fire into space. The unicorn turned away from the red sun toward the dark void of deep space. The sun shrank into a tiny red dot and disappeared behind the Milky Way. They came near a belt of small rocks and asteroids. She saw two large space stations hiding behind two small planetoids. Both space stations lit up as she flew near. She could see busy construction robots all over them. After a while, Victoria and the unicorn approached a large planet. She recognized it as Jupiter with the atmospheric storms and the Great Red Spot. Earth would not be far.

Victoria was glad to be home, but she saw something disturbing. Swarms of space vehicles appeared to be firing at each other causing some vehicles to explode into smithereens. She thought, "OMG, unicorn, evade, evade." But it was too late; she felt a powerful pulse hit her as she fell off her chair back in Huang Shi's room, inside an underground bunker somewhere near Anyang.

The noise woke Huang Shi, who abruptly sat up in his bed. He looked dazed and puzzled. "Where am I? Who are you?" said Huang Shi in Putonghua, which sounded like "wo-zai-nar?" (我在那儿; I am where?) and "ni-shi-shui?" (你是谁; you are who?)

Victoria regretted she never learned Chinese.

"Sorry, I don't understand," she said excitedly, "let me find someone to translate."

"No need for that," said Huang Shi. "I speak English. I'm asking where am I and who are you?"

"You are at a safe place near Anyang," replied Victoria. "I am the baby you gave to your son Huang Di to look after."

"The baby was a girl. You look like a boy," said Huang Shi. "And you're big. How many years have passed? Where is Didi?"

"I am a girl. My name is Victoria. I'm thirteen years old," said Victoria. "Your son Didi disguised me to look like a boy to confuse the Watchers. But they trapped us, and they zapped your son with a bright light from their briefcase. He vanished after that. I'm sorry."

Huang Shi sat there expressionless for a while. At length he spoke stoically, "It's important that you're safe. Princess Victoria, please lock the door and make sure no one can see or hear us. We need to talk in private."

Victoria locked the door and disabled the surveillance cameras. She was not fearful of being inside a hole in the ground in the middle of nowhere. The frail man in front of her had travelled through time and space. Nothing could stop Victoria once she learned how to do the same.

"I must have awakened because of you. You are very powerful," said Huang Shi. "I'll tell you everything I know about your parents, and then you'll have to make a dangerous trip to save them. You'll need them when the Oracle comes true. I remember your parents as if it was yesterday. They were young, brave, and in love. They came from two royal households, both alike in dignity; from ancient grudge, the lovers rose in mutiny …"

Chapter 52

The One-Eyed Fable-Monger

The two rebel lovers from the mountains finally reached their destination. They were in dire straits and desperately in need of help.

"Young man, I sense great joyful celebrations but beware of disaster," the blind fortune-teller hollered at Wen Long and Xin Feng as they walked past. "Come sit down; let the expectant mother have a rest. We have a family special, and I'll throw in a free couplet for the baby."

It took a long time for Wen Long and Xin Feng to arrive at Anyang. They walked, they climbed, they waded, and they raced. They worked odd jobs, stole food, told sob stories, and hitchhiked. They kept moving, making it hard for their clans to catch them. But something more ominous was on their tracks. Wen Long and Xin Feng kept seeing two men dressed in dark jackets, one of them very tall, the other carrying a briefcase, always showing up just a few days after the young couple had arrived in a town or city. When that happened, they quickly moved on to the next destination.

The trek to Anyang took months. Along the way, Xin Feng got pregnant. It wasn't something they planned. In fact, very few Haryan had reproduced sexually since they got used to the *in vitro* fertilization and *ex vivo* electronic surrogacy technology. When the classic signs of Haryan pregnancy started to show, Wen Long and Xin Feng began to worry. Even with significant genetic disguises to make the Haryan on earth resemble humans, they did have a number of anatomical and physiological differences.

Xin Feng stopped ovulating, fat began to accumulate in her abdominal area, a tentacle that would secrete milk grew out of her navel, her nails turned red, her hair turned silvery, and her skin sparkled under the sun. This condition would last for one month until the baby was born. The newborn baby would be about the size of a thumb. It would stay in a gestation pouch at the mother's belly where it fed on the secretions from the milk tentacle. It would take four to five months for the baby to grow about seven to nine pounds. Afterwards, the baby would be weaned and grow just like a normal human baby.

Xin Feng tucked her hair inside a wide brim hat with a veil and stayed away from the sun so people wouldn't notice the sparkle. The one-month period was almost up. She was about to go into labour any day now. Xin Feng sat down at the fortune-teller's desk. She needed a rest. The baby drew a lot of energy and nutrients from the mother. Xin Feng felt hungry, thirsty, and tired.

Although this was deep winter, the sun was out, there was no wind, and the temperature was up. Even the street fortune-teller set up shop to do business.

"How do you know my lover is with child?" asked Wen Long as he sat down. In order to be in tune with the times, Wen Long used the progressive term 'lover' rather than the traditional 'wife' or 'spouse.'

The blind seer lifted his dark glasses to show the young couple that his right eye was fine.

"In the land of the blind," said the fortune-teller with a twinkle in his good eye, "the one-eyed man is king."

On the fortune-teller's desk were several sheets of scrap paper held in place by a yellow stone. There was a fortune-teller's almanac with a tortoise shell drawing on the front cover. Three ancient words ran down the centre of the shell, which any Chinese could read as 'thousand mile path' (千里路). It was an esoteric term for the vertical line that bisected the tortoise plastron (abdominal plate). Only high priests, augurs, and professional fortune-tellers knew the meaning of this term.

The half-blind seer took a sheet of scrap paper and said, "I feel generous today. I'll give you a family consultation for only one yuan. Satisfaction guaranteed or your money back."

"Since you are a fortune-teller, you should know that we don't have money," said Wen Long. "But I need some help to find a place for my lover to rest in private. Just for a few days. I also need to get her some food and drinks. I will work and pay you back."

"Stop calling me a fortune-teller. People frown on this profession nowadays," said the one-eyed man. "I'm a fable-monger. As for your predicament, I may be able to help. Do you have any IDs?"

Both Wen Long and Xin Feng shook their heads.

"Please, we are honest people," Wen Long said. "I'll pay you back. I promise."

"Alright, I'll be a fool," said the fable-monger. "I'll do it, but I need some collateral."

"All we've got," said Wen Long, "are just the clothes on our backs."

"That is not true," said the fable-monger. "You may think that I'm a fool, but only a fool without vision can see. And I can see that both of you have pendants hanging around your necks. Just leave them with me, and I'll get a room for you and your lover. When I get paid, you get your pendants back."

Wen Long and Xin Feng looked at each other. They really had no choice. Wen Long had an idea. If these royal trumperies weren't of much use to him and Xin Feng, they might as well sell them.

The fable-monger however didn't want to buy the pendants but he would send the young couple to a jeweller who might. He folded a sheet of scrap paper and gave it to Wen Long. "Take this to Master Huang the jeweller-watchmaker around the corner. Give him this note. He'll help you. Go."

When Wen Long and Xin Feng turned the corner, they saw a man in a small stall with a loupe on his eye, working intently on an antique watch. Wen Long made his presence known. The watchmaker let down the loupe and lifted his head. He was also blind in the left eye. He was none other than the fable-monger himself. Both Wen Long and Xin Feng were startled.

"That's a good trick," exclaimed Wen Long. "How did you do that?"

"Do what? Fixing the watch?" asked the bemused watchmaker. "It's my profession. I've been doing this for thirty years. What can I do you for?"

"Do you have a twin brother," asked Xin Feng, "who is a fortune-teller?"

"No, I don't have a twin brother," said the watchmaker. "And if I did, he won't be telling any fortunes. The Party is very much against superstition, and so am I."

"Well, a fortune-teller who called himself a fable-monger gave me this note to give to you," said Wen Long. "You're Master Huang, right?"

"Let me see," said the watchmaker, unfolding the note as he peeked at the young couple from the corner of his eye.

"*Stone man with one eye* sent you," said the watchmaker. "Well, why didn't you say so? You should've said so right away. I'm at your service. What can I do for you?"

"We want to sell our pendants," said Wen Long. "Will you take a look and make us an offer?"

Wen Long placed both *huang* jade pendants in the watchmaker's palm. The watchmaker put his loupe on his left eye and scrutinized the pendants for a full minute while Wen Long and Xin Feng nervously waited for his verdict.

The watchmaker let down his loupe and pushed the pendants back in Wen Long's hand. "I have a problem with these," said the one-eyed watchmaker who just turned a blind eye to rare treasures. "These two pieces are priceless to you, but worthless to everyone else. Since the *Stone man with one eye* sent you, I'm going to help you out. Keep going to the end of the block and turn right. You'll see an apothecary just around the corner. Give him this note, and he'll look after you. Go."

Wen Long and Xin Feng did as they were told. Turning the corner, they saw a small Chinese medicinal herb pharmacy with a wall of small drawers containing all kinds of rare herbs. The apothecary was checking the stock and had his back facing the door.

"Excuse us," said Wen Long. "We have a message for you."

The apothecary turned around. He had a patch over his left eye, and he looked like a dead ringer of the fable-monger and the watchmaker. They must be triplets.

"How do you do that?" exclaimed Wen Long. "Are you an illusionist?"

"What utter nonsense!" the apothecary said, "I have been a *daifu* (大夫; great expert, Chinese term for doctor) and an apothecary in the neighbourhood for almost thirty years. You can ask anyone in the neighbourhood. They all know me."

Just then, a young girl of about ten passed by; she was followed by a young puppy.

"Xiao Bei," hollered the apothecary, "Do you know me?"

"Of course I know you," Xiao Bei hollered back. "You're our favourite Huang Daifu. Do you need me to get you something from the market?"

"Are we happy?" the apothecary asked Wen Long, as he waved off Xiao Bei. "Now what can I do you for?"

"We need help," said Wen Long, handing the note to the apothecary.

"*Stone man with one eye* sent you. Well, why didn't you say so?" said the apothecary after reading the note. "You should've said so right away. Come on in."

Wen Long and Xin Feng looked at each other, both having the feeling of déjà vu at the same time.

"Let's see now," said the apothecary. "It says here you want to sell your pendants. Can I have a look?"

Wen Long put the pendants in the apothecary's palm.

"They are priceless to you but worthless to the rest of humanity," said the apothecary after thorough inspection. "Since the *Stone man with one eye* sent you, I'll take the pendants as collateral. Follow me."

The apothecary pocketed the pendants without waiting for an answer from Wen Long or Xin Feng. He opened a small door and led them into an unlit room with a narrow flight of stairs. "Walk this way," said the apothecary as he went up, taking two steps at a time. Although from the outside, the shop looked like a single level shed, the steps seemed to keep going up for several stories. At the top was a door, which opened into a large room. The window view showed that they were no longer in Anyang but somewhere in the countryside near a mountain.

This mysterious one-eyed man was obviously a great illusionist. A younger man was in the room. He immediately helped Wen Long put Xin Feng in bed and started boiling a kettle of hot water on the stove to make tea. He was preparing dumplings in another pot. Wen Long thought the whole set-up rather staged, but he was thankful someone was helping him take care of Xin Feng.

"He is my assistant," said the apothecary. "Just call him Shazi (傻子; idiot). He will be here to help you with whatever you need. He may be an idiot, but he is not a moron. I have work to do. Everything will be fine."

As he headed out, the one-eyed apothecary stopped at the door to share a bit of news with Wen Long.

"By the way," he said, "two men wearing black coats have been prowling the neighbourhood looking for a young couple. Make sure you avoid them. I know you have many questions. Shazi will explain as best he can, but we can't tell you everything because it is against the law. Good luck."

CHAPTER 53

The Dragon's Dream

Xin Feng had a good meal. She fell asleep after drinking several cups of a fragrant flower tea that Shazi brewed. Shazi reassured Wen Long that it's just Amaranth tea, perfectly safe for Xin Feng's condition.

"Congratulations, the baby princess will be born in the morning," said Shazi, who could tell the sex of the baby by the classic signs, such as the colour of the mother's nails turning red rather than blue. "She will enter the gestation pouch by herself at sunrise. It will be painless for the mother. The Amaranth tea will keep her relaxed so that there will be no stress for both the mother and the baby. Have you thought of a name?"

"We agreed that if it's a girl," said Wen Long, "we'd call her Wenxin, taken from our names."

"The heart of literature sculpts the dragon, that's a good name," said Shazi. "You should rest too. Tomorrow is Winter Festival. The sun will stop moving south and start moving north again. It's a day of transition. It's going to be a big day for everyone."

Wen Long slept on the floor after Shazi left. It was the softest floor he had ever slept on. He also had the sweetest dream. Wen Long dreamed that he and Xin Feng would somehow survive everything and be together again in the end with their baby. He would find a place for his family to live and thrive in. They would have more children and the children would in turn have children. On the weekends or the holidays, the children would visit and dine together. He and Xin Feng would grow old together surrounded by millions of descendants.

Where is this paradise? Is it on earth or is it only in his dreams? Where all parents are strong and wise and capable, and all the children are happy and beloved. Where brothers and sisters neither slaughter each other, nor enslave one another, nor tell lies to everyone from morning till night. Where the children love and help each other, share everything and take care of one another, and all of them live peaceful, fruitful, and happy lives. Where is this place? Is it China?

Xin Feng was already up and about when Wen Long woke from his beautiful dream. She had given herself and the baby a hot bath. The sparkle on Xin Feng's face was gone; in its place was a glow. Her silvery hair was changing colour. It was now blonde to reddish and would soon return to her natural dark brown. The red colour of her nails had already faded. The Thumbelina sized baby Wenxin was safely born. Wen Long felt great joy admiring this larval life form while she busily suckled on her mom's lactating tentacle inside the gestation pouch. She would feed voraciously and grow in the pouch for the next four to five months.

After eating the Chinese fried choux known as *youtiao* (油条; oil strip), hard-boiled eggs, and multi-grain congee for breakfast—Shazi had bought them breakfast and left—Wen Long and Xin Feng discussed their situation.

"The House of Sheru built our civilization based on Hugin—thought," said Wen Long. "We have developed mind-reading powers. I can tell that Huang Daifu and Shazi have no evil intent. The House of Sachem built your civilization based on Munin—memory and we are here in Anyang because you have memory of this place. Someone had planted memory of Anyang into your family's genes so that we would end up here. I think everything is happening as it should."

Just then, Shazi entered the room. He had a black briefcase with him.

"It's okay, I'm not one of them," said Shazi when he saw the couple react. "Huang Daifu decided against his better judgment to lend us this briefcase and for today only. This is what created the space warp that discombobulated you yesterday and now hiding you here where the Watchers can't find you."

"I know about space warps," said Xin Feng. "My clan has developed a crude version of it."

"Where is Huang Daifu?" asked Wen Long, "When will he be back?"

"Huang Daifu has gone off on one of his unpredictable journeys," said Shazi. "No worries though. I have instructions from him, and he has left your pendants here."

Shazi sat down with Wen Long and Xin Feng at the table. He first gave them the note that was passed from the fable-monger to the apothecary the day before. Wen Long and Xin Feng looked carefully at the scrap paper this time and realized that it was a page from an academic article. There was a small map inset showing the Huan River (or Anyang River) and Yinxu. South of the river in the middle of the palace area of the Shang ruins was a penciled cross, and beside it the words '8 m.'

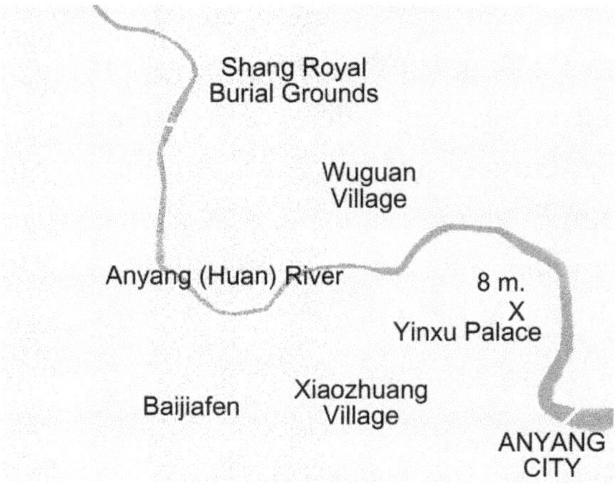

The author of the article was Huang Shi. Wen Long flipped the page over and found these hand-written words: "The sun stops going south, the oracle is here; find the yellowish stone, trust it with all your heart; use the map to locate, a treasure very rare; the ring of unicorn, leave in the baby's care; follow the moron's lead, whither be here or there." *Stone man with one eye*, Dec. 22, 1975.

Huang Daifu didn't leave any explanations, so they tried to decipher the message themselves. The sun stops going south—that would be the winter solstice. The oracle is here—the oracle has come true. Find the yellowish stone—find the author of the article Huang Shi, which means yellow stone. Trust it with all your heart—trust Huang Shi with Wenxin (Victoria), the Heart of Literature. Use the map to locate, a treasure very rare—the map inset shows the exact location where a rare treasure is buried. The ring of unicorn, leave in the baby's care—the rare treasure would be a ring with a unicorn motif; the baby should keep the ring. Follow the moron's lead, whither be here or there—always follow Shazi's advice, in whatever direction he points.

"But how do we find Huang Shi?" asked Wen Long.

"One of the contributors to Huang Shi's paper is the head of the Institute of Archaeology, the famous Professor Zheng," said Shazi. "She'll open the doors for us."

"What makes you think," asked the earthbound prince, "that Professor Zheng will give us the time of day?"

"By promising a unique discovery of the ages," said Shazi, pointing at the cross on the map, "Prince Wen Long, you can be very convincing if you use more of your mental powers. It's a gift from your ancestors. Just speak firmly to Professor Zheng like this, 'You will tell Huang Shi to come immediately.' Professor Zheng will give you Huang Shi, and she will find our unicorn ring for us."

Shazi then gave the young couple two pieces of black fabric folded into the size of a handkerchief. "These will unfold into a large sheet," he said. "When the Watchers banish you to temporal exile, wrap yourself in this fabric and cover your head. It will save your life."

Finally, Shazi took a small box from his coat pocket. It had the drawing of a plant on the cover ().

"Here is a small present from Huang Daifu to the baby princess," said Shazi. Inside the box were the two royal *huang* jade pendant pieces tied by intricate knots with a red string to make a necklace. "Seeing how you two will not be able to pay for your room and board, Huang Daifu has decided to take possession of the collateral and give it as a present to the baby princess. This must stay with her at all times. When Huang Shi takes the baby to safety, I will be gone, and where I go, I cannot take you. After I leave, you will lose the protection of the space warp. I suggest you two split up to confuse the Watchers. It will buy everyone more time. But the Watchers will eventually catch up to you. Let us hope your daughter will grow up safely and come rescue you. Our future depends on it."

"What's this?" Xin Feng asked when she saw the Oracle Bone words on the pendants.

"Huang Daifu added these glyphs," said Shazi. "It is the Oracle of the Apocalypse. The prophecy of three thousand years has come true for the secret clans. The princess will need a lot of help before she is ready. The future depends on how things unfold at Anyang."

Xin Feng was getting very upset about the prospects of losing her daughter and her lover. Wen Long put his arms around Xin Feng to comfort her. Since that day the young lovers met in the remote mountains, much water had passed under the bridge. The bond between them had grown strong and they had become inseparable.

"Have courage, my love," said Wen Long. "All these will blow over. I have been having very good dreams about our future. I'll go now with Mr. Shazi to find Professor Zheng and Huang Shi. You can stay here with the baby and design a secret mark for her. That way we will always recognize our daughter."

Fast forward to early May of 1976, Wen Long showed up as promised and indicated the exact location of where Professor Zheng should drill. They quickly reached the foundation of the ancient Shang palace. Normally, that was the time to stop. No one had ever found anything under Shang palace foundations in the past except virgin soil.

Wen Long told Professor Zheng to keep drilling. He insisted with a polite but firm voice, "Rare treasures are buried deep. You will keep on drilling."

Professor Zheng wisely decided to comply. Somewhere around eight meters or twenty-six feet below ground level, Professor Zheng and China struck gold. Small but priceless artifacts started surfacing. Soon it became unmistakable to everyone present that they had discovered something major down there.

One of the first artifacts that came up was a jade ring with a mythical beast motif. Somehow, it ended up in Princess Wenxin's treasure box, tied with the two *huang* pendant pieces into a necklace. In the chaos of the early excitement, Shazi's replacement ring easily fooled the handlers. That night Shazi came to the apartment for the last time.

"This is it. My job is done here," said Shazi. "The unicorn ring is only one of a pair. In time, they will reunite. The words at the back of this ring say 'go' and 'come,' meaning 'peripatetic.' That is the name of the unicorn. Princess Xin Feng, do you remember how to pronounce the word peripatetic in Haryan language?"

"It's Galizu," replied Xin Feng.

"You will tell Huang Shi the name," said Shazi. "It is through this name your daughter will one day invoke and command the unicorn to come and go in time and space. When your daughter grows up, she will have special powers. There is much you do not know, but let us keep it that way. We must never violate the Uncertainty Principle or causality, for that'll be the end of us."

"Will we meet again?" asked Wen Long. "Huang Daifu and you have helped us so much. We hope to repay you someday."

"There are many reasons why we help you, not all of it altruistic," said Shazi. "You'll learn in due time, when the inexorable current of time takes you to port according to schedule. I think somewhere along the journey, we will meet again. We do not know when or where because the future hasn't happened yet. Good luck to you and to us all, Godspeed, au revoir, auf Wiedersehne, adiós, sayanara, and zaijian (再见; again see)."

Fu Hao tomb at Yinxu

CHAPTER 54

The Long March

WARNING: Graphic description of the apocalypse within. Handle with care.

Back in the underground bunker at the present time-line, Huang Shi told Victoria about the last time he saw her parents in Mid-May of 1976. The young couple was sitting across the dining table in his apartment. The father was distraught and disheveled. The mother's face was as pale as a sheet. In her arms was a pretty baby wrapped in a red blanket. This was the first time Huang Shi saw the baby princess he was supposed to protect. He was expecting her to be a bit larger at five months old.

"My lover and I must leave town immediately," said Wen Long. "The Watchers have been combing the streets of Anyang looking for us. You should take our baby and leave without delay. It is not safe for any of us."

Xin Feng buried her face in Wen Long's embrace. She could not bear the thought of losing her baby, but it must be done. Huang Shi took the princess in his arms. She was sleeping peacefully with no idea of what was going on. Wen Long passed the treasure box to Huang Shi, with the instruction for how to invoke the unicorn Galizu. After that, Wen Long had to drag the tearful Xin Feng out of the apartment.

Huang Shi sat on his bed with a baby in his arms contemplating what he had gotten himself into. Why did they choose him to carry the burden of human existence? Huang Shi had seen the end of the world from the *Virtual Quantum State Simulator.* That experience traumatized him.

Heaven turned into a death-marked shroud, while Louis Armstrong sang the dirges of the Wonderful World. The earth was covered to the horizon with corpses, piled up one upon another to make mountains of carrion. There was no one left to bury the dead.

Streams of blackened blood carried the severed limbs and heads into tributaries that joined the raging River Styx. Some of the bodies were cleaved into pieces, while others were burned and charred into macabre and monstrous shapes. The tiny bodies of the children were blackened and bloated by poison and starvation, their vacant glassy eyes filled Huang Shi's nightmares.

Even maggots and worms refused to visit this feast of offal offered by this hellish abattoir—this infernal slaughterhouse we call earth. The monsters that caused this did not enjoy impunity after all, for they and their children lay among the soulless dead.

Huang Shi could hear only the icy wind weeping and wailing, making feeble and empty protestations to the cruel gods. Hell was indeed freezing cold as Dante had described. Humanity's crime condemned Earth to suffer a million years of nuclear winter. Huang Shi must make sure the baby princess would grow up safely to forestall the approaching apocalypse. He must steel himself to overcome the direst situation. Even when all hope was lost, he must never give up.

As Huang Shi started packing, he thought about his own life. He remembered how he survived the tragedies of his parents and his wife, and how he made it through the lost decade. Now he was burdened with the knowledge of the world's end, and the duty to protect a newborn alien baby who would one day save humanity from extinction. What deranged moronic imbecile would believe in such a lie? Huang Shi could hardly convince himself this was not a cruel joke or a bad dream. He didn't really know if he had the mental strength to do what was asked of him. He needed to understand the meaning of never giving up.

What does it mean to never give up? Huang Shi remembered a news story about an airplane crashing in the Andes a few years back on a Black Friday (Uruguayan Flight 571, Friday, Oct. 13, 1972). The survivors of the crash found themselves in the middle of the magnificent mountain surrounded by snow, with no food, no heat, no warm clothing, no electricity, no radio, no mountain climbing equipment, and no idea of location. They were also in the middle of bad weather and in danger of being buried alive by avalanches.

By any measure, these Uruguayan youths were already dead, or as the saying goes, "As dead as Shrödinger's cat." No one had given them so much as a snowball's chance in hell to get out of there alive, unless the snowball was at the ninth circle of Dante's inferno, where it's freezing cold.

On the eleventh day, the dying transistor radio delivered the news that the authorities had called off the search and rescue. Every sane rational person in the civilized world had given up. Fortunately, some of the survivors did not give up. A dying boy Gustavo said, "… ***good*** news … they've called off the search." Someone shouted angrily, "Why the hell is that good news?" Gustavo would soon die from the ordeal, but he saved many of the survivors by his answer, "Because it means that we're going to get out of here on our own."

They formulated a plan—a mad plan but the only one that would keep them alive. Two of the survivors Nando and Roberto would eventually do the impossible. They walked out of the mountains on their own and saved the remaining survivors after seventy-two harrowing days in the snowy mountains. That was never giving up.

Huang Shi told Victoria how he feared having to face the trials and tribulations that would soon descend upon him. He needed to know how mentally strong people faced extreme difficulties. He thought about the birth of the Chinese Communist Party.

The once proud Empire of the Dragon was humiliated by wave after wave of foreign incursions at the end of the Qing dynasty. The fall of the Qing dynasty and the establishment of the Republic of China (ROC) brought a ray of hope, but the country was promptly balkanized by warlords and carved up by foreign colonial powers. The ROC, founded by American educated Dr. Sun Yat-sen and based on the American model, waited anxiously for support from the West, which never materialized. The Western powers gravitated towards the warlords instead. It was more profitable to sell arms than sell democracy and human rights, and it was cheaper to buy a warlord than to win the hearts of the Chinese peasants. It was easier to enslave and plunder China if its people were divided and weak.

China, as an ally of the victors of World War I, had sent a hundred and fifty thousand labourers to the battlefields in Europe. Eleven thousand Chinese men gave their lives for the common cause. Their homeless spirits still wander the poppy fields of France and Flanders. The Treaty of Versailles however transferred the previous German colony in Shandong province to Japan. Liberal President Woodrow Wilson decided that his own Fourteen Points and Chinese aspirations be damned. It was more important to placate Japanese demands. Realpolitik trumped the Magna Carter of the soul.

Thus, China was betrayed. True friends stab you in the front. China was the only ally that did not sign the treaty. For people who asked three decades later the question "Who lost China?" They might look for their answer here. They might also discover "Who lost Vietnam?" Ho Chi Minh, the key founder of the Socialist

Republic of Vietnam, couldn't get past the door to the conference. A wise man once said, "If you're not at the table, you're on the menu." Ho knew then that the West didn't dole out free lunches. His people must pay with blood for their freedom from their colonial masters.

The betrayal of China by their Western allies touched off the May 4th Movement led by intellectuals and students. Some of China's scholars decided that they needed to find an alternative route to independence. They would no longer depend on the vain hope that the moralizing and pontificating West would grant China and its people self-determination out of righteousness. It was good news that the West took off the mask. For some Chinese revolutionaries, they realized that they would have to walk out of the mountains on their own.

The Bolshevik Revolution in Russia became a living example of a successful struggle of the common people against imperialist powers. In China, a small group of book-toting academics armed with mighty pens decided to start their own Communist Party. In the absence of their two founders, thirteen academics convened the First Congress of the Chinese Communist Party (CCP) in 1921 in Shanghai. They hoped communism would unite the Chinese people to fight for the unification and independence of China, without which social and economic reforms would not be possible.

Most of these early communists were not entirely clear what the new ideology meant, or whether this European idea of the proletariat revolution would work for China. There were also scores if not hundreds of such organizations that were sprouting up all over China in those heady days of revolution. Who knew which one would survive in the end?

Out of this initial group of thirteen members, plus the co-founders who were not present, only two survived to attend the inaugural ceremony of the People's Republic of China (PRC) on

October 1st, 1949 at Tiananmen Square; they were Mao Zedong (毛泽东) and Dong Biwu (董必武). All the others had left the Party, joined the Nationalists, sold out to the Japanese, got excommunicated, or died engaging in the dangerous profession of being a communist revolutionary. The odds for survival were 13%.

Until the end of 1948, the ruling class of China and most from the West, including the Soviet Union, never expected the Chinese Communists to be able to survive let alone succeed. At the beginning of 1948, even the Chinese Communists optimistically estimated that the Civil War would last another ten years.

All of the early Chinese Communists were academics who had never wielded a weapon or had any experience in war. They hardly knew how to hold a proper fist. Mao Zedong was a peasant's son who rose in status to become a librarian's assistant for crying out loud. They were up against the warlords and the Nationalist army led by Chiang Kai-shek, their erstwhile brother-in-arms, which the colonial powers, the landed gentry, and crime bosses supported with manpower, money, equipment, and guns. What chances did the Chinese Communists have? Yet they did not give up.

After receiving no help from the West, Dr. Sun Yat-sen's Nationalist political party Kuomintang (KMT) worked in cooperation with the CCP and received material support from the Soviet Union. At that time, the ROC government had a precarious existence. Its rule was limited only to the province of Guangdong (广东; Canton) in the south, and even that was not a sure thing. The warlord who invited Dr. Sun to set up his government in Guangdong turned around and bombed Dr. Sun.

With the help of the Soviet Union, the ROC raised and trained an army with the expressed objective of eliminating the warlords and reuniting China under one flag—the ROC flag. Dr. Sun however died in 1925 without achieving his China dream. Thereafter, the ROC government and the political party KMT fell under the control of the military led by Chiang Kai-shek on the conservative wing.

After the purge of the communists by Chiang Kai-shek in the April 12th Incident in 1927, during which, Chiang's henchmen ambushed and slaughtered a large number of communists and labourers, the survivors of the CCP dispersed into the countryside. Captured communists had the choice of publicly renouncing the CCP and selling out their comrades, or suffering torture and receiving summary execution. Mao's young wife had the same choice when she was arrested, but she chose death.

The CCP leadership, many of whom were also concurrent KMT members and ROC executives, woke up one morning and found their faces on Chiang's Dead-or-Alive Wanted posters. It was China's version of *Night of the Long Knives*. A true friend stabs you in the front, but a brother riddles you with a shower of bullets from all directions and then throws in a live hand grenade for good measure.

The top leader of the CCP, Chen Duxiu (陈独秀), took the blame and was excommunicated by the party he co-founded. A warlord captured and hanged the other co-founder Li Dazhao (李大钊). Betrayal, torture, and execution were waiting for the early communists at every turn. The CCP was in disarray, and it did not have soldiers or arms to defend its right to exist. What odds did the Chinese Communists have? Yet they refused to give up. Huang Shi didn't understand what drove those young men and women to lay down their lives just to dream the impossible dream.

The CCP did not curl up and die. Mao escaped to the remote mountains of Jinggangshan (井冈山; Well Ridge Mountain) in Jiangxi (江西; River West) province to set up a peasant soviet. The fledging revolution movement had many existential problems. Mao lost his leadership position at the CCP to a clique of Soviet educated youngsters known as the 28 Bolsheviks, which precipitated the battlefield disasters for the First Red Army in the Nationalist's *Fifth Encirclement Campaign*.

The Chinese Communists, with a little more than a hundred thousand ragtag peasant soldiers, were hopelessly surrounded by over a million well-armed, well-equipped, well-trained, well-fed, well-funded, well-fortified, and well-led Nationalist soldiers. In addition, Chiang Kai-shek had German military experts advising him on how to tighten the noose and strangle the bastards. For Chiang, it would be like making bugsplats. What were the odds of survival for the Chinese Communists? Couldn't they tell it was hopeless? Yet they refused to give up. To save themselves, they walked out of the mountains on their own.

Thus started the legendary 25,000 *Li* (里; Chinese mile) or 10,000 kilometre (6,000 mile) *Long March*. During this arduous journey, the indomitable Chinese Communists broke through encirclements, fought heavily defended Nationalist positions, crossed rivers that were not meant to be crossed, climbed insurmountable peaks, marched through forbidding terrain, sometimes under extreme conditions, with pursuing armies hot on their heels, and Nationalist planes raining bombs whenever weather permitted. All the while, this band on the run was beset by starvation, thirst, sickness, wounds, weariness, and despair. Dying was easy. Almost every other day there was a crisis that threatened to destroy the CCP and its so-called army.

In their darkest hour, the CCP decided they would follow Mao the ex-librarian assistant and part-time poet whither he wanted to take them, even if he led them across the Red River (赤水河; chi-shui-he, red water river) four times not telling anyone in which direction they would be heading next.

Then near the end of the trek, one of their own, in fact one of the thirteen founding members, Zhang Guotao (张国焘), almost rent the CCP asunder and threatened to cause a civil war within a civil war. Zhang's staff Ye Jianying (叶剑英) arguably saved the CCP by warning Mao of Zhang's intentions to take power by a military coup. Mao escaped by the skin of his teeth. Years later, Ye Jianying would support the rise of Deng Xiaoping against the *Gang of Four*. Marshal Ye did the CCP and the Chinese nation a couple of big favours, and he was not the only one.

Out of about eighty-six thousand souls at the beginning of the Long March, less than ten-percent arrived one year later at the remote northern city of Yan'an (延安; extended peace) three hundred kilometres (180 miles) north of Xi'an, and that's with fresh recruits along the way.

If England's Dunkirk Evacuation, Hannibal's Crossing of the Alps, and Xenophon's Anabasis of the Ten Thousand seem herculean, the *Long March* of the Chinese Communists is at least ten times maybe a hundred times more so. It is hard to find any other example of a military operation that suffers the loss of over ninety percent of the personnel and does not crumble into a pile of dust.

China has an idiom for extremely risky undertakings: "Nine will die; only one survives." The *Long March* was such an undertaking. Why would anyone of sound mind take such bad risks in return for the chance to implement some farfetched revolutionary idea that may or may not work? Despite the insanely unfavourable odds, the Chinese Communists never gave up.

Huang Shi explained to Victoria how he was immediately tested.

"I heard your mother's voice asking me to open the door. I thought she forgot something," said Huang Shi. "I opened the door and found two men wearing dark shirts standing in front of me. One man was a giant, and the other carried a briefcase. I had you in my arms with nowhere to run."

CHAPTER 55

Invictus

When the remnants of the Red Army arrived at Yan'an by the end of 1935, it was a small group of underage, undernourished, unkempt, unshod, and unarmed peasants masquerading as soldiers. On the other hand, they emerged from the crucible that forged each soldier's body and mind into unyielding steel. The soldiers of the *Long March* were so tough they would forget to run for cover during enemy air raids. For them, bombs raining from the sky were no more bothersome than a light drizzle at a garden party.

The CCP's troubles were far from over. They immediately faced a war of extermination against the Nationalist's Northwest Army led by General Yang Hucheng (杨虎城) and the Northeast Army led by General Zhang Xueliang (张学良). Both armies were well known fighting forces led by famous generals. Their numerical superiority over the Chinese Communists was at least twenty to one. What chances did these crazy communists have? Did they really believe they could survive? Why didn't they just give up and go home?

The Chinese Communists didn't give up. China was their home. Instead, they convinced the two Nationalist generals that their common enemy was the encroaching Japanese Empire, which had already occupied northern China without any resistance from the Nationalist army. This precipitated the Xi'an Incident in which the generals turned around and arrested Chiang Kai-shek. That was when the CCP leader Zhou Enlai stepped in and negotiated for Chiang Kai-shek's release. What ironies!

For his troubles, General Yang Hucheng and his whole family, including his wife and young children, would be knifed to death by Chiang Kai-shek's agents in Chongqing (重庆, also known as Chungking, China's wartime capital). General Zhang Xueliang, Manchuria's warlord who had voluntarily joined Chiang's Nationalist cause, ended up living under house arrest in Taiwan for over fifty years until the death of Chiang Kai-shek's son. Mao Zedong and the CCP meanwhile had an eleventh hour reprieve. The CCP's peasant armies joined Chiang Kai-shek's Nationalist armies to fight the Japanese occupation in China.

After the end of the Second World War, the Chinese Civil War restarted. The war against Japan swelled the ranks of the Chinese Communist army, but the Nationalist army had three and a half times as many soldiers. Moreover, the U.S. armed and trained Chiang Kai-shek's Nationalist troops, which had an endless supply of American dollars, lethal rifles, deadly machine guns, heavy artillery, unstoppable tanks, mechanized transports, unlimited ammo and fuel, railways, planes, and warships, led by seasoned generals. The Chinese Communists had the weapons they captured from their enemies. Their leader was a librarian and a poet. Even Stalin hedged his bets by signing a treaty with Chiang's ROC government. Except for the support of the peasants, the Chinese Communists were on their own.

Chiang Kai-shek estimated it would take him about six months to destroy the CCP. It was no empty boast. The Nationalist's Northwest Army now under General Hu Zongnan (胡宗南) promptly captured Yan'an, the capital of the CCP, sending Mao and his men scurrying off into the countryside again. What made the crazy Chinese Communists think that they could defeat

Chiang's powerful armies in conventional warfare? Why wouldn't they just give up and go home?

"Give us the baby and go home," said the towering 7'6" giant, who was about as tall as Yao Ming, the former all-star player of the Houston Rockets. "This is way over your head. The baby will be safe with us."

The shorter stocky guy with a flat nose was blocking the doorway. Huang Shi concluded that he could not escape, but giving up the baby was not an option.

"We are messengers of Dias the Infinite," said the giant. "We come in peace. Please don't force us to use violence."

"I think we should just grab the baby," the flat-nosed one at the door said. "This guy is a stooge. He doesn't know Jack about what he's involved in."

Huang Shi looked at the cute little baby with her big bright eyes. What could he do other than give up? He could try to invoke Galizu. It was a crazy idea, but he couldn't give up.

The Chinese Communist did not give up either. Instead, they went on to annihilate the Nationalist armies. Only fourteen years after the *Long March*, they took over the job of governing five hundred million long-suffering peasants in a war-torn country.

The newly established Communist government had no money. Chiang had taken the nation's gold reserve as well as the national historical treasures to Taiwan. There was no international political recognition. The Soviet Union boycotted UN in protest when its Western members refused to recognize the PRC.

Soon an embargo would block Communist China from trading with most Western countries and their dependencies. It was a global trade war designed to bring China to its knees. Meanwhile, large numbers of lawless bandits, organized remnants of Nationalist soldiers, and Chiang's secret agents continued to roam the country, creating havoc whenever and wherever they could.

The Constitution of ROC considered all of China as its legitimate realm and Chiang Kai-shek threatened to reconquer mainland China from Taiwan.

The new Communist government had to convince the warlords of the remote regions to lay down their arms. They had to diplomatically or otherwise ask belligerent foreign soldiers and their gunships to vacate Chinese territory. Furthermore, the Chinese Civil War would not end until Taiwan's reunification with the Chinese mainland. Just when the People's Liberation Army (PLA) was amassing along the Fujian (福建; or Fukien) coast preparing for the final battle, the Korean War broke out. I. F. Stone had asked the question *cui bono*, or who benefits, from this eruption of hostilities on the Korean peninsula. The surprising answer was Chiang Kai-shek.

When the Korean War broke out in June of 1950, the PRC government was still a newborn babe doing its best to survive under difficult and adverse conditions. It could not permit American soldiers to establish itself across the Yalu River (鸭绿江), where Chiang Kai-shek with U.S. backing could open up another front to invade the Chinese mainland. Mao didn't want the Korean War but the gods of war would not cooperate. His repeated warnings to General MacArthur not to cross the 38th Parallel fell on deaf ears. No one, least of all MacArthur, would entertain the remotest possibility that the despised Chinese peasants would dare to challenge the almighty American slaughter machine.

But American bombs were already falling on Chinese villages. Mao had no choice but to send his war-weary and recently disbanded peasant soldiers into North Korea during deep winter to battle the most lethal fighting forces the world had ever known.

The U.S. led coalition army had complete control of the air. They had already flattened the entire Korean peninsula and had run out of targets. The U.S. side had heavy weapons, artillery, tanks, helicopters, and mechanized transports. Their soldiers had warm clothing, heavy boots, unlimited ammo, hot food, and MASH (Mobile Army Surgical Hospital). They could easily move troops and heavy equipment around with their air transports and warships.

The U.S. and coalition armies were led by the famous five-star general Douglas MacArthur, who proposed to use scores of atomic bombs on Korea and China, turning the whole region into a radioactive wasteland. The massive deaths of the yellow skin natives, who were allies only five years ago, didn't even figure in the calculations of these gods of war. That was how the Western civilization proposed to pacify the five thousand year old Chinese and Korean civilizations in Asia. We have a lot of Nobel prizes to thank for this warlike, bloodthirsty, and hubristic mindset.

The Chinese Communists blithely entered the war during the frigid cold of Korean winter with their scantily dressed, primitively armed, and poorly supplied peasant soldiers, who travelled by foot each carrying nothing more than eight days' ration of rice. They could only move at night and in the mountains where they wouldn't be bombed to oblivion by the USAF.

Mao's eldest son Mao Anying (毛岸英), who was imprisoned at the age of eight and forced to witness his mother's execution, who survived as street waifs in Shanghai with his two younger brothers, and who later fought the Nazis with the Soviet Red Army as an American ally, gave his life at the Korean front. Were the Chinese Communists crazy? What was Mao thinking, sending his excellent son and heir apparent into the line of fire? What made these primitive peasants imagine that they would have any chance fighting against the greatest slaughter machine in the history of humanity?

As the Watchers walked over to wrest the baby from Huang Shi, they slowed down. Time is a strange thing. It's all relative. Maybe Huang Shi sped up; or maybe it was his winged thoughts that accelerated. Huang Shi's mind plunged into the baby's eyes. It was a dark space with a tiny flicker of a flame. As Huang Shi came near the flame, it began to dance wildly and burn with vigour.

"Galizu!" thought Huang Shi, and the thought echoed in his head. The flame flared up into a fireball as the unicorn emerged. It was a beautiful and magnificent animal with large bright eyes. Its sparkling golden skin was cool in the flame. Huang Shi knew that it was the baby that evoked the unicorn. Her mind was somehow linked to him, and she was responding to his wishes. Galizu came over to lick the face of the baby, making her giggle. Unicorns love babies because they're innocent and incapable of deception.

"Galizu, take us to my home at Yellow County," said Huang Shi in his mind. "Take us one month into the future so the Watchers won't be able to track us."

No sooner did Huang Shi have that thought when his body went up in flames. Huang Shi blacked out. He entered a dreamland where he saw Wen Long riding a horse along a river. The young prince was dressed in skins and had a beard. His long hair was braided. Another man rode up beside Wen Long. This person had a banner of the scorpion billowing in the wind. More men rode up. They were all armed with bows, sabres, and spears.

Wen Long pointed at the river with his spear and announced, "Shar Moron." Then he pointed the spear forward and shouted, "Dalai." The horde roared in approval and charged forward like a thundering tsunami, sweeping past all obstacles before them.

Time did not make sense in dreams. Huang Shi felt that he had a one-eyed visitor who spent a thousand years whispering wisdoms in his ears. The mystery man's last words were, "Time to wake up."

Huang Shi woke in an unfamiliar neighbourhood. He felt as if all his bones were disjointed. It wasoeep winter but all the snow had melted beside him. The heat under his skin made him feel like he was going to burst into flames. He felt something moving in his arms. The baby was safe. She was still giggling from the unicorn's kiss.

Huang Shi was lying on the ground just inside the alley beside his home. He knew where he was but he couldn't recognize the neighbourhood. Multistory apartments and brightly lit shops occupied the fields. Even his home was completely renovated. The whole neighbourhood had electric light, telephone, and colour television. There were even cars parked on the street. It was already after dark. Everyone was inside their heated homes, feasting on dumplings, fish, chicken, and pork, downing prodigious amounts of high alcoholic content *baijiu* (Chinese vodka) with their kith and kin at the year's most important meal, the Winter Festival dinner. The new world in front of Huang Shi was a dazzling display of opulence.

"My God," Huang Shi thought. "What future is this that Galizu has taken us?"

CHAPTER 56

Quantum Leap

Although China did not want to be involved in the Korean War, once the Western armies began arriving at the Yalu River, China had no choice but to respond. Everyone in China ate less food in order to send their peasant soldiers into Korea. Chairman Mao's most excellent son sacrificed his life. Others in China did the same. To the imperious General MacArthur's surprise, the despised Chinese laundrymen pushed the mighty American army back to the 38th Parallel.

General Joseph Stilwell, popularly nicknamed Vinegar Joe, finally received vindication four years after his death. He had maintained that the Chinese peasants could be good soldiers if properly trained, armed, and led. He just didn't reckon it would be the Chinese Communist peasants who would vindicate him, and against the American juggernaut to boot.

The impossible became reality because the Chinese Communists never gave up. Independence and the right to be responsible for one's own actions are not gained by rabid barking or through the master's largesse, but by blood and sacrifice, lest anyone forgets.

Victoria was mesmerized by Huang Shi's story. But if she ever faced an unbeatable foe, would she have such an indomitable will?

"All over the world," Huang Shi said, "there are millions and millions of people who live in despair, usually through no fault of their own. They are just born in the wrong place at the wrong time under the wrong circumstances. Some are taught to accept their lot in life, others are told to find their rewards after death, and still others find the only way to survive is to give up their human dignity. Some will even sell their soul.

"Most of these unfortunate people live a wretched life, sometimes oppressed by those who have been oppressed in a vicious cycle of endless atrocities and injustice justified by a world of lies, ignorance, prejudice, and cruelty. And when the hopelessly wretched comes to the end of their short and tortured life, they ask not why they are dying but why they have ever lived. What is the point of it all, just to live and suffer?

"The apocalypse will set the world on fire. Earth and her children will suffer greatly. The Oracle says that you are the one who will save the world. When darkness descends, it will be up to you to rid the world of despair, and to teach the people how to walk out of the mountains on their own. It is most important that you understand what it means to never give up, for one of these days you will be tested."

Victoria exuded intrepid calm while Huang Shi made fiery prophesies of fire and fury. She remembered the harmless interpretation of the stopping-of the-sun.

"Put the unicorn ring on your thumb, Princess Victoria," said Huang Shi. "Your real odyssey starts now. This ring makes you the master of a unicorn called Galizu. Just call for the unicorn by name, tell it where and when you want to go, and you'll get there. You must find your real parents and bring them back to the present time-line. They will prepare you for the final battle. But you must remember never to break the *Law of Causality.*"

"Do you know when and where my parents are?"

"I had a vision that your father Prince Wen Long lived as a nomad near a river by the name of Shar Moron," said Huang Shi. "The river goes to a place known as Dalai. His tribe uses the sign of the scorpion on their banner, the same symbol on your *huang* jade. Galizu will read your mind and locate your father at the synchronized time and place. I do not know where and when the Watchers sent your mother Princess Xin Feng, but I'm guessing it has something to do with Fu Hao of Shang. This archer-ring that invokes Galizu came from Fu Hao's tomb. Go now. I'll be fine."

When Victoria invoked the unicorn's name, the light in the room dimmed to complete darkness. From faraway, almost at infinity, a tiny glow burst into flames. The distance between Victoria and the flame narrowed. The flame grew bigger and started to dance wildly. Victoria remembered this from long ago. "Galizu!" thought Victoria, and the magnificent unicorn emerged from the fireball.

When Victoria woke up from the shock of her first experience of quantum leap, she could see that a few boys and girls were examining her weird clothing with great curiosity. A friendly puppy was sniffing her but did not dare to get too close because of the heat. Blue steam was rising from her clothes and her skin. One boy was rummaging through her knapsack, which Victoria carried with her wherever she went. The sound of metal clanking came from a nearby smithy busily forging weapons.

It was freezing cold, like Toronto in winter. Victoria felt feverish from the heat generated during her bi-millennial trip. When she came to, she realized that the unicorn ring was not on her thumb. Someone had stolen her ring. In that moment of confusion, Victoria garbled, "Where is my unicorn?" Victoria knew that if she lost the ring and couldn't invoke Galizu, she would be stranded forever in a strange place in time. All she had gone through would be for naught.

Victoria's self-healing was powerful. The discomfort of time travel quickly dissipated. She got up with a groan and grabbed the kid who had her knapsack, “Hey, stop it. Where is my ring?” Victoria pointed to her thumb.

The kids immediately dashed off in all directions. Victoria felt at a loss when she heard a shriek from the direction of the stream behind the smithy. She ran over and found a boy screaming in pain while thrashing in the stream. His skin must be red hot because he was shrouded in sizzling steam, like hot steel being quenched by cold water. Victoria ran into the stream and took hold of the kid. Her healing powers immediately soothed him. She retrieved the ring and put it securely back on her left thumb. The unicorn ring knew its mistress.

Some grown men and women had shown up to find out what the ruckus was all about. They began to yell at Victoria in heated language, thinking this alien in strange garb was harming their children. Several men were running over with long sticks and the blacksmith emerged from his forge with a sword. Things were going down schiesse spiral very quickly. Victoria had to think on her feet. She remembered David’s advice about using magic when surrounded by hostile savages. Victoria held up her hands showing her open palms to the excited audience. She rubbed her palms together a few times, and showed them again, this time with a drawing of the scorpion on each palm.

The excited folks immediately calmed down. They formed a circle around the stranger and argued whether they should take the boy to the Tuman or skewer him first and then take the body to the Tuman. A horseman rode over and the circle promptly opened up for him. He pointed his spear menacingly at Victoria and jabbered some gibberish. Victoria showed him the clan symbol on her palms. The horseman backed off as well. He made the sign ‘follow me,’ and then the whole gang paraded all agog towards the Tuman’s tent.

Victoria took the chance to check out the surroundings. The smithy was on the outskirts of a large walled settlement. It was early morning. People were out and about herding sheep, goats, oxen, and horses. There were reindeer pulling large wagons loaded with goods. Everyone seemed to have a Siberian husky or a hunting hound. The houses were clean and sturdy, organized in neat rows. There were inns, food stalls, and shops all along the street. This was not barbaric at all.

The parade through the settlement was drawing a huge crowd, and everyone was interested to find out about this short-haired strangely dressed unisex creature that seemed to have dropped in from outer space. Soon more horsemen joined the parade. Messengers galloped off shouting for people to stay out of the way.

What Victoria didn't expect was to see a huge camp hidden in a gentle valley beyond the village. There must have been thousands of men in the camp. Something was afoot. One could easily tell there were many different tribes present because of the different banners. In the centre of the camp was a large tent with the scorpion flag. As with most military camps, a wide lane led directly from the front gates to the chief commander's Tiger Tent.

Victoria's heart was racing like a colt out of control. Would she finally meet her real father, an alien from outer space? What would that make her? What did the aliens originally look like? Did they have a big sucker in the middle of their face with sharp fangs all around it? Did they have slimy tentacles? She never had a chance to think about it. She had really enjoyed being a normal human. She wished that it were all a nightmare or a joke. On the other hand, she was strong like Supergirl, invulnerable like Wolverine, and could travel through time like Marty McFly. Most people wouldn't mind being her. Maybe she could enjoy the benefits of being an alien superhero with a human face.

CHAPTER 57

The Name of the Flower

The person sitting in the chief's throne was older than Victoria had imagined. Although his face wasn't wrinkled and creased, his hair and beard was grey and white. Something was wrong. If Galizu synchronized everyone's relative time, her father should be in his thirties. The man came down from his throne to approach her. He looked carefully at her. Took her hand with the thumb ring and checked it over. Then in Putonghua, he said, "deng le ni hen jiu a, nü er."

Poor Victoria knew it was her father, but there was a generation gap, language barrier, and cultural shock between them. She shook her head and shrugged with a helpless smile, "Sorry, I don't understand."

Fortunately, her father was a most capable alien who could speak twenty earth languages including Coptic, Aramaic, Farsi, Turkish, Armenian, Amharic, Urdu, Hindi, Gaelic, Arabic, and Russian.

"That's okay, I understand you," said Wen Long, "I was saying, what took you so long? You're my daughter Wenxin. I'm your father. Your mother had given you secret markings so we can recognize you whenever you show up. Come here and give your father a hug."

Victoria had long suspected that her 'heart' tattoo wasn't her only ID; her real secret identity marking was hidden in plain sight.

Victoria guessed it while learning about her background from David. Her father had similar markings.

"Where is my mother?" asked Victoria after enjoying the fatherly embrace. She could almost feel the joy that was emanating from her long lost father. "Is mommy here with you?"

"No, I don't know exactly where or when the Watchers sent her, but I have an idea," said Wen Long. "After entrusting you with Mr. Huang Shi, I separated from your mother. We promised each other we would try our best to survive until you showed up, and then we would look for each other, no matter where, no matter when."

Father and daughter caught up with the lost years. Victoria finally learned where she came from, and where human civilization originated. She quickly felt very comfortable and close to her father, as if she could hear his thoughts. She had inherited her father's mental powers. They could practically talk without speaking.

Wen Long had a theory of why Galizu sent Huang Shi and Victoria twenty-eight years to the future instead of one month.

"Galizu came from a mythical planet known as Shangala," said Wen Long. "The month and years at Shangala must have been different from earth's cycles, hence the extra years. Now it's too late to change that."

"It's not too late," said Victoria, "I can call up Galizu and go back in time to correct the mistakes."

"No, that will violate causality," said Wen Long, "I have lived my years; I cannot un-live it. You have collapsed my quantum state relative to yours. Let us find your mother at the right time-line and then we can figure out together what to do next."

"I don't understand," said the puzzled Victoria. "Aren't we in the past? Haven't we changed the past?"

"It's not exactly the same when we're talking about relative quantum states," Wen Long said. "I'll explain later, but first we'll have to find your mom, and then you'll take us to the present timeline. You are the only person who can save us from temporal exile. But before we go, I have a few matters to settle.

"You're seeing here a gathering of the banners. Tomorrow, we decide on migration routes for various tribes that want to expand beyond our home grounds. These tribes have a great future. Someday, their descendants will rule China and influence the people and culture there. Even though they will disappear as a unique people, their blood will be a part of China's blood, and their stories will be a part of China's stories. They just won't be called Xianbei, but what's in a name?"

"I know. Mr. Huang Shi's son told me the story of Xianbei," said Victoria. "I didn't know you have a hand in it."

"I didn't do it all by myself," said David. "As a chief, I had a helpmate. But she disappeared in the mountains almost twenty years ago, shortly after your stepsister Blitzen Wolf was born. We never found her body. Blitzen Wolf and her husband Wolfman have come under my banner, as I have no male successor. You will meet them. After that, we'll go."

"I thought aliens and humans don't mix," said Victoria.

"Blitzen Wolf was procreated by expert gene splicing," said Wen Long. "A doctor and his assistant did that using future technologies. I met the men two thousand years in the future. I didn't realize I would meet them again in the past."

The tribe held a big feast that evening. Victoria was dressed in the fashion of a tribal princess, with a foxtail hat, an embroidered vest, deer skin pants, sheepskin boots, a black sable coat, silver and jade bracelets, bejewelled anklets, and a gem studded dagger, etc. Victoria's half-sister Blitzen Wolf, who was similarly attired, joined her at the table. She had a red flower painted on her forehead; it's a flower that Victoria thought looked strangely familiar. Lone Wolf gave Blitzen the idea. He based it on the Mark of Diamaranth. This would one day become a popular Xianbei and Tang fashion.

The language barrier unfortunately limited the conversation between the sisters to hand gestures meaning eat, drink, yummy, and can't take another bite. It didn't matter to Victoria. She was exceedingly happy to know that she was not all alone in the universe. Victoria wished that she could use her iPad to take a *Thelma & Louise*, also popularly known as a selfie, with her older sister, but she remembered the causality problem and held back.

Singing and dancing filled the evening. Victoria did the *Lambeth Walk* from the English musical *Me and My Girl* to everyone's delight, and she got half the tribe to follow her on the shuffle dance.

What happened to Wen Long aka Lone Wolf and his tribe after settling in the mountain range? As he had told his tribe, the Hunnu turned their attention westward and southward. Wen Long knew his Chinese history.

Meanwhile, more and more remnants of the various surviving Eastern Wolf tribes came through the mountains and joined Lone Wolf, who provided them with food, lodgings, and safety. Other groups that escaped the slaughter had settled in the mountain range to the north and south of Lone Wolf, forming kindred alliances and helping each other defend against the Hunnu or the natives. Lone Wolf reorganized the tribes under different banners, allotted land, taught the people to supplement hunting with fishing and farming, and built well-defended settlements.

Lone Wolf became an expert in breeding animals, a skill he passed on to his daughter Blitzen Wolf and her husband Wolfman. In a few years, the banners were gorged with animals, especially sheep, horses, and reindeer. The reindeer was a part of the myth that saved the tribe. It became the fabled unicorn called *Zhi* (廌) by Han Chinese but known as *Shanbe* by the Eastern Wolf tribes. In order to make everyone feel that they're one family, Lone Wolf created a super tribe, took the name of *Shanbe*, and adopted the unicorn as its totem. The mountain that had protected them and nurtured them, they also named *Shanbe*, later known as the Greater Khingan Range (大兴安岭; da-xing-an-ling). Three hundred years later, when *Shanbe* tribesmen started showing up at the doorsteps of the Eastern Han Empire, the Han Chinese gave the tribe a Chinese name, Xianbei (鲜卑).

The *Shanbe* expanded so fast that with hardly a fight, it took over all the land from the Berserkers, which probably felt that it was to their advantage to become a part of the *Shanbe* super tribe. By this time, *Shanbe* had become the dominant force from the mountain to the plains along the minor yellow river known locally as the Shar Moron, or Xar Moron, literally 'yellow river' in the local language. It's a small stream from the mountain, which joined with others to become the major Liao River in Manchuria, where the descendants of the Xianbei known as the Khitans later built the Liao Dynasty. Many nations still use the name Khitan and Khitai to mean Chinese and China.

The Liao, river of iron, mother of the Khitans, debouched through the Dragon's Mouth into the vast limitless *dalai*. 'Dalai' means ocean in the local language.

It was the destiny of the three hundred horsemen with their women and children who survived the slaughter of the Huns, to build a future empire that would rule China and become a part of China; just like how the Shar Moron becomes a part of the ocean. There is an old saying in China: "The Sea can accept a hundred rivers."

The tribesmen elected Lone Wolf Tuman even though he didn't want the title. What's in a name? White Wolf and North

Wolf decided to part ways with Lone Wolf. They took a core group with them down south. They thought that Lone Wolf was too timid. The wealth was in the south, where the land was rich, the weather was warm, the beds were soft, the men were effete, and the women were pretty. Once they left the Shar Moron, they joined with southern groups and called themselves Wolf-Huns (乌桓: wu-huan).

The timid *Shanbe* stayed out of trouble and grew strong slowly. The Wolf-Huns meanwhile went south into Han territory. Some say they established their home base in Manchuria at the White Wolf Mountain (白狼山: bai-lang-shan), named after their leader White Wolf. They survived many years triangulating between Han and Hun. As fate would have it, the mighty Wolf-Huns met their ignominious end in the decisive Battle of White Wolf Mountain.

There was no record of how many were slaughtered, as the counting would be too arduous, and the number would not be believed. But the survivors who surrendered stood at more than two hundred thousand. The famous anti-hero Cao Cao (曹操) of the *Romance of the Three Kingdoms* utterly destroyed the Wolf-Huns and erased them from the face of the earth. The few that refused to die or surrender formed a new tribe known as Xi (奚), eventually to become an integral part of the Khitan Liao dynasty.

After the conference of the banners concluded, Lone Wolf called a family gathering with Blitzen Wolf and her husband Chieftain Wolfman. They were very successful animal breeders. Aside from magnificent war steeds and racehorses, they had bred huge cows and woolly sheep. As for man's best friend, Blitzen had bred about thirty different dog breeds, some for hunting, some for herding, some for guarding, and some only for companionship. There were all sizes and shapes, some cuddly cute, while others deadly dangerous.

The most incredible animal they successfully bred was the unicorn, which was actually a miniature reindeer with a single horn. They called it the Luduan (甪端).

Lone Wolf asked Wolfman and Blitzen Wolf to hold the fort while he had to be away for a while. He had left orders to the generals, clan chiefs, and the elders to seek the leadership of Wolfman in his absence, and after three years, if he did not return, to let the tribe elders select a new leader, or if they so choose, to continue growing under Wolfman. He left his clan banner, the Tuman spear, as well as his valuable belongings with Wolfman. Blitzen Wolf and Wolfman didn't think anything was out of the ordinary because Lone Wolf had previously taken solitary retreats deep in the mountains for months at a time.

Before departing, Lone Wolf told Blitzen Wolf's children the story of the tribe, which would one day become a part of China. Dynasties would rise and fall, but *Shanbe* would always be there, among a larger group known by its name as an elegant flower, Hua. What's in a name? That which we call a flower by any other name would smell as sweet.

Lone Wolf and Victoria went deep into the mountain to make their journey through time, going back a thousand years into the past to the early reign of King Wu Ding.

Chapter 58

Grand Unification

Lady Diane had just shot an owl with a harpoon arrow. Normally, she would not hunt owls because Queen Zia loved them and kept them as pets. Lady Diane had no choice; she needed to eat so she would have the energy to keep climbing. She had been running from the soldiers of King Wotan for the past eleven days. She was hiding in the mountain about eighty kilometres (fifty miles) to the northwest of the capital. The gorges here were steep and treacherous.

Lady Diane knew about this place because this was where the famous Red Flag Canal (红旗渠: hong-qi-qu) would be built during the 1960s. The canal is an engineering wonder completely built by hand along the cliffs of a mountain and through scores of tunnels, carrying water for sixty-five kilometres (forty miles) from Shaanxi (陕西) province to irrigate the fields in He'nan. She had visited the site with her lover Wen Long, that brave boy from the *Heart of the Dragon* when they were together in that time and space. That year was like a dream, full of life, love, adventure, and sunshine, even though they were always hungry and constantly on the run. Love has a way of making suffering feel like pleasure.

While she was eating the roasted bird, Lady Diane looked at the archer-ring on her thumb, that special personal gift from Queen Zia. When Queen Zia let Lady Diane choose from two identical rings, she almost took the wrong one. If she had taken the Galizu ring, then she would have no choice but to surrender herself to be killed and buried with Queen Zia. How else could Professor Zheng unearth the ring from Fu Hao's tomb, which allowed Shazi to steal it, letting it end up in Wenxin's treasure box with the flower symbol? Causality would have been a bitter pill to swallow.

Lady Diane learned that Queen Zia had needlessly bled to death from the bungled delivery of her fifth child. The whole royal midwifery team was therefore sacrificed. They were then buried with the Queen to be her slaves in the afterlife. Lady Diane was at that time pacifying another minor tribe. It was most unfortunate. Her presence at the delivery would have saved many lives. On the other hand, saving lives that should not have lived was a dangerous thing to do.

At her deathbed, Queen Zia asked for Lady Diane, her dearest friend, and someone whom she had grown to love. The broken-hearted King Wotan however interpreted that as the queen's request for Lady Diane to follow her into the afterlife.

Lady Diane heard a noise in the woods and immediately stood up with her bow drawn. To her surprise, two barbarians emerged—one was a man with greying hair and the other a young girl. She dropped the bow and uttered, "My god, what happened to you?" Of course, to Victoria, it sounded like "wo de tian na, ni zen me la?"

Fortunately, Victoria's mother spoke twenty-five languages including Zulu, Gujarati, Betawi, Tagalog, and Akkadian. Victoria thought when she returned to Canada, she would learn French and Chinese as a start, maybe Spanish, Italian, and German on the side, even Latin, Greek, and Hebrew. But Akkadian?

After the tearful kisses and embraces, the family settled down and exchanged stories of what happened to each other during all these years they were apart. Wen Long remarked that although all three of them were separated by thousands of years, they were linked by objects and events that eventually brought them back together. Victoria thought that her real parents were totally awesome, and she felt that together they could achieve anything.

Victoria and her mom compared their rings. Victoria's ring had the words: () meaning 'to go' (去: qu) and 'to come' (来: lai). Xin Feng's ring had the inscriptions: (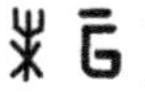) meaning 'return' (回: hui) and 'not yet' (未: wei).

When Xin Feng put her ring on top of Victoria's ring, then read from top to bottom and from right to left in the Chinese manner, the four words became Return (回), Go (去), Not Yet (未), Come (来). In Chinese, 'return go' (回去) means 'go back to,' while 'not yet come' (未来) means 'the future.' The four words read this way meant '***Back to the Future***.' This didn't mean anything to Wen Long and Xin Feng, but Victoria knew that the person who named the two unicorns and created the rings of power was most certainly a space-time traveller, and no doubt a humorous one.

"You should keep the rings," Xin Feng said, as she gave Victoria both rings. "But never use their power without careful consideration about the damages caused by time anomalies. There is always a catch and a price to pay."

"Do you know anything about the Oracle of the Apocalypse?" asked Victoria. "According to Mr. Huang Shi's son, it may be just a scary term for the winter solstice. But Mr. Huang Shi was very serious about it, and said that I would save the world. If I have a choice, I would rather live a simple life as a normal girl."

"We're not sure what the Oracle is all about," replied Wen Long. "Quite frankly, we don't believe in prophesies. According to the Uncertainty Principle, the future hasn't happened, and no matter how clever we are, we cannot predict the future to occur with absolute certainty. We most certainly cannot travel there. There is no there there. We should not believe in fortune-tellers and fable-mongers."

"You mean all this apocalyptic oracle mumble-jumble," Victoria exclaimed, "could be just a bunch of baloney?"

"We don't know why someone has gone through such tremendous troubles to create the secret clans," said Xin Feng. "We still don't know who the *Stone man with one eye* is and how he figures in everything. At the end of the day, the Oracle motivated the secret clans, it kept you from the Watchers, and it has led you to us. It is a mystery but we are thankful."

"Your mother and I were supposed to die in temporal exile, never to see each other or you again," said Wen Long. "Now we are reunited as a family and we will be the first to escape temporal exile. I have dreamed it and now it's coming true. Let us be happy with how things have turned out. We can worry about the rest later."

Victoria was very relieved to hear that the oracle was bogus, and that no one could predict an apocalypse to occur for certain. Perhaps she didn't have to save the world after all. Although Victoria missed Michael and Angela, she now looked forward to starting a new life with her super-cool biological parents, one of whom was a tribal chief and the other an ancient warrior. Maybe they could work on a plan to rescue Michael and Angela from time-exile.

But while the family was chatting excitedly about the past and the future, they forgot about the present. The Shang soldiers had stealthily crept to within fifty metres from them. With a signal, the soldiers started running at full tilt with their nets, lassos, and long poles ready to capture their prey. It all happened so suddenly and at such speed that Victoria and her parents had no time to prepare for time travel. They had to react in the traditional manner. They turned tail and ran as fast as their legs could take them.

Soon they came to the end of the road with a sheer drop of a hundred metres (three hundred feet) in front and nowhere else to go. The soldiers were about to nab them in a matter of seconds. Victoria couldn't believe that the rescue of her parents from time-exile would end in abject failure. How could they be so careless?

"Girls, hang on to me and jump," yelled Lone Wolf. "Don't worry, I have Plan B."

Victoria closed her eyes and made the leap of faith. She had to trust her father. She remembered the time when she jumped out of an airplane. That time, she knew that she was attached to David who had a tandem parachute. This time, she could only hear the wind whistle in her ears. When she opened her eyes, she saw her mother smiling at her with not a modicum of concern. It was the look of total trust on her face. Victoria heard the cry of an eagle, and then she felt a gust of wind as the whole family soared into the sky.

Wen Long's black cloak was a present from Shazi. Xin Feng also had one. The cloak was made of a memory fabric that could turn into a glider. Wen Long had tested it when he went into the Greater Khingan Range by himself.

The wind that hit the face of the cliff created an updraft that lifted the glider into the air. The pursuing soldiers looked on in wonderment at how a black bird, the totem of Shang, grabbed Lady Diane and flew into the heavens. It was a sign that she was a goddess. The soldiers prostrated as one on the ground and worshipped their gods, while the recently reunited royal family of Shangala flew off into the setting sun.

CHAPTER 59

Jack in the Box

Victoria's disappearance from Huang Shi's room in the bunker gave everyone a fright. Huang Shi had fallen back into a coma and couldn't offer any information. After a couple of days of frantic but fruitless search, Charlie and Viola got a call from Victor. They would find out that Victor was Victoria, and that she needed Uncle Xu to get Huang Shi over to Canada. Huang Shi needed to be near Victoria in order to begin repairing the damages he had suffered from his time anomaly.

Charlie and Viola took the family shopping for modern clothing and accessories. It was Charlie's turn to gawk at the vivacious Victoria. He was speechless the whole time. Victoria had great fun teasing Charlie. Obviously, she was attracted to him as well. Charlie was smart, fun, and quite an Adonis himself. Even the Weird Sisters accepted Charlie and Viola into their midst. When they parted, Charlie promised to visit Victoria in Toronto when—not if—he was accepted at Princeton.

Without getting into mundane details, the *Weird Sisters* and their respective families returned to Toronto, and tried to resume life with some semblance of normalcy. Wen Long and Xin Feng both got English names. Wen Long became Piero while Xin Feng became Luna.

One day, Victoria took her parents to the mall while worrying aloud that their money would soon run out.

"Don't worry. When we look around us," said Piero, "we see everywhere signs of people who are related to us. When you need help, always go to family first."

"You mean alien relatives," asked the bemused Victoria, "from other planets?"

"We're talking about the Shang people," said Luna. "They are everywhere, hiding in plain sight, and some have become very powerful. Daddy, you're a Shang, why don't you explain?"

This is the story of the Shang people after the collapse of their dynasty. When the Zhou tribe took the mandate of heaven from the last Shang king Di Xin (帝辛; Theo VIII), they allowed the royal family of Shang to continue as vassal states.

The Zhou sent one branch under Jizi (箕子), the uncle of Di Xin, to rule the northeast region. Koreans call him Gija, and consider him a founding father of the Korean civilization. Jizi likely came from the branch of Yarjee (亚其; ya-qi), the father of Fu Hao. Another branch under Wu Geng (武庚; Martial VII) son of Di Xin stayed to rule the Shang population at the old capital of Yin (殷).

Wu Geng however joined a rebellion and died as a result. The Shang capital Yin was abandoned. It would become the Yin Ruins, or in Chinese, Yinxu (殷墟). Thereafter, the Zhou sent Weizi (微子), the elder brother of Di Xin, to the ancestral Shang city of Shangqiu (Shang mound), where he ruled as the Duke of the Song State (宋), keeping alive the royal lineage of Shang under another name.

Weizi is the ancestor of many major Chinese clan names including Song (宋; old spelling: Soong of the Soong sisters), Hua-flower (花; of Hua Mulan), and Hua-China (华; of Hua Guofeng, Mao's successor). Now we know where the Shang royal lineage has been hiding. Weizi's younger brother Weizhong (微仲) succeeded him as the second Duke of Song. Confucius is the fifteenth generation direct descendant of Weizhong. Two thousand and five hundred years of Chinese culture stood on the unshakable foundation of a Shang royal descendant. Is this a coincidence?

Outside of the Song State and the Shang aristocracy, the ordinary Shang subjects had a tough time. The Zhou dispersed them and banned them from working as scholars, court officials, soldiers, or peasants. The only job they could do was the lowliest profession known as trading. Can you imagine a lower social status than peasants? That was to be called a Shang person—a merchant-trader.

The first famous and historically significant Shang trader was a person by the name of Lü Buwei (吕不韦). Is it a coincidence that he was born in Anyang? Lü Buwei invested his entire family fortunes in a person, the father of Qin Shi Huang (秦始皇). Because of this lowly Shang trader's bold investments, Qin Shi Huang would one day become the lord of the Qin state and later the emperor of the first unified Chinese empire. Without this Shang person Lü Buwei's intervention, China might never have become a unified country using a unified language to form a unified culture. China would be like Europe. Is this a coincidence?

In time, the term Shang person (商人; shang-ren) became the regular expression for a merchant-trader. Nowadays, whenever a Chinese say merchant or businessperson, they're actually saying 'Shang person,' but they don't realize its original meaning. That is how the Shang descendants hide in plain sight.

Furthermore, business in Chinese language is 'Shang enterprise' (商业; shang-ye); a company trademark is a 'Shang mark' (商标; shang-biao); a shop is a 'Shang store' (商店; shang-dian); merchandise is 'Shang products' (商品; shang-pin), and so on and so forth. Most people don't realize all these business terms originated from the Shang tribe.

"When we say we see Shang people (商人; shang-ren) everywhere," said Piero, "your mom and I are saying we see merchants and business people everywhere, many of them are very successful and they contribute positively to the economy and society. Even some of the heads of the most powerful states in the world are Shang people."

"Some Shang descendants have been quietly helping us," said Luna. "They are hidden in plain sight. For example, that treasure box with the *huang* jade pendant has a drawing of the flower bush on the lid. It is the Oracle Bone word for flower, and it has become the word to mean China. The drawing is also the Shanglyph that represents the Diamaranth, the immortal flower of Shangala, the paradise planet. The Diamaranth is a real flower known only to people like us. I have seen its real shape as a global company's logo. I bet we have blood kin working in this company."

"You mean we may have rich cousins? Yay!" exclaimed Victoria. "Which company is that?"

"It's a Shang enterprise from China known as Huawei," said Luna. "Its logo is a drawing of the real Diamaranth flower."

It suddenly dawned on Victoria why the flower on Blitzen's forehead and at the centre of the star chart in Michael's journal looked familiar. It was the Huawei flower logo.

"Do you know what Huawei means?" asked Piero.

"No, I've always wondered," said Victoria. "Huawei is quite famous. Their ads say 'Wow Way.' Is that what Huawei means?"

Victoria's parents went on to reveal the secret to the clues that had been staring at her all this time. The flower logo of Huawei is the representation of the first word of their name *Hua* (Trad: 華; Simp: 华). It is the word which originated from the Oracle Bone symbol for flower (), developed into the word that means elegant, and later becoming the word that means China. To identify someone as a Chinese is to call him or her a Hua person (华人; hua-ren); namely a flower person, or an elegant person.

The second word *wei* (Trad: 爲; Simp: 为) of Huawei is the Chinese word meaning to be, to do, to act, or to make. It is a most important word. The drawing on the inside of the treasure box is the Oracle Bone symbol for *wei.* It shows a hand controlling an elephant (). This is however not Shanglyph from Shangala, but was created by a scribe from the Shang tribe who developed the word from experience. Did the Shang use elephants to do their heavy lifting, or did they get that idea from their Promethean ancestors, who migrated all along countries where people used elephants to do work? In any case, the two symbols on the treasure box says, '*Hua*—China, and *Wei*—to make.' Whoever put the words on the box was trying to tell us a message.

"We suspect that someone important in the company is Shang royalty," said Piero, "probably a descendant of Weizi or Weizhong, and very likely related to a visitor from the stars."

"And we shouldn't be surprised that Huawei named their most advanced System-on-a-Chip the Kirin," said Luna, "which in Chinese is pronounced Qilin, the Chinese unicorn. Who else besides the Haryan believes in unicorns?"

"Huawei is inviting suggestions for their promotional slogan," said Piero. "This is my message to our secret kin in the company: *Stone man with one eye* submits this slogan, 'Huawei—China Works!'"

Here we are at the end of Victoria's odyssey. We would like to tie up a few loose ends before the curtain falls.

The *Stone man with one eye* has popped up all over the place but no one can confirm who he is. We do know that King Kull cultivated the symbol of the 'one eye' to carry special meanings. The Shanglyph of the eye means to be, to do, to make. The 'one eye' symbol has survived in the ancient Egyptian Hieroglyph

(𓁹), which unsurprisingly meant to do, to make. It is also the name of Osiris (𓊨𓁹𓀭 ; the eye is the pronunciation, the throne is a divine symbol, and the seated figure is the god determinative), the god related to Orion. The chief Egyptian falcon god Horus, the son of Osiris, also has an important eye symbol, the all-seeing *Eye of Horus* (𓂀). The eye, the black bird, Orion, and the name Horus are all related to the Haryan and King Kull. Are these all coincidences?

In China, while 'a hand controlling an elephant' has replaced the eye to mean 'to do' or 'to make,' the important 'one eye' symbol has not disappeared. David called himself the one-eyed servant because his ancestor Huang Yin had used this title. Huang Yin was the one-eyed trader who gave the archer-rings to Queen Zia. He called himself the King's Eye. That word in Oracle Bone script is a humble downcast eye (臣: chen; OB: ). Its original meaning was slave or servant, later taken to mean official of the court, and a self-deprecatory address when speaking to the king. An official carries the connotation of someone who acts.

The 'one eye' survives to this day as a secret symbol hidden in plain sight. Check out the extraordinary one-eyed people in history, such as Hannibal, Lord Nelson, Theodore Roosevelt, and the Communist general Liu Bocheng (刘伯承), nicknamed the War-god. Don't forget the Norse god Odin and the one-eyed cyclops Polyphemus in the Odyssey. How about James Joyce the author of Ulysses who wore an eye-patch. There are numerous one-eyed characters in movies such as Harry Hart in Kingsman, Elle Driver in Kill Bill, Ruth Buggs in Paul, and Nick Fury of the Marvel Universe. The list is endless. Even the Ellen Show has a One-eyed Monster. Of course, the American dollar has the all-seeing eye design, which is a respected symbol of the Masons. The world also fears the evil eye.

Let us go through every thread and check for loose ends. Every one of Victoria's perilous adventures described in the first chapter has come to pass. We have described all the incredible and improbable events with no exaggeration. We have reported the facts faithfully; no more, no less, no lie, as requested by Victoria herself.

Finally, this is a peaceful tale; we want to reassure our readers that no one died unnecessarily. The rest are bugsplat and collateral damage. Are we sure Raja died? No one ever reported his disappearance and death. No one ever found Wolf Star's body either. What happened to her?

We confirm that Michael, Angela, and David survived but woke up in unknown times and places in the past. Victoria and her parents will try to find them and rescue them. But time is immense; they will need clues. Victoria will one day get help from an awesome magician by the name of David Blaine. He has clues that David Huang ended up in dynastic Egypt as Pharaoh Khufu's magician, Michael woke up during Persian Emperor Xerxes's reign to become the right hand man of the female warrior Artemisia, and Angela languished in Byzantium while building the influential royal lineage of the Angelos from scratch.

We will not be able to answer all questions. Do we know how Victoria's parents recognized her? What is the rare treasure in Michael's journal? Does Blitzen Wolf know she has alien genes? Who is Santos Didymas? But the most maddening puzzle of all—why does the number eleven dominate events and mysteries in our relative quantum reality? The answers are hidden in plain sight, but only a fool without vision can see.

For Victoria's world, peace has returned. Peace always returns. Everyone is exceedingly happy. All's well that ends well, except when the end is a catastrophe.

CHAPTER 60

Apocalypse

"And I saw the seven angels … and to them were given seven trumpets … and there fell a great star from heaven, burning as it were a lamp … And the name of the star is called Wormwood." – Revelation Chapter 8

The Day of Destiny had arrived. This was the last game of the Global Major League Baseball postseason, the seventh and deciding game of the New World Series. The War on Halloween caused worldwide social disruption and extended the last day of the postseason to the 11th day of November, a day during which Alibaba broke all records again with sales of RMB 1.1 trillion yuan for the day.

The Toronto Blue Jays of the Liberty League had fought tooth and nail to claw its way here, outperforming everyone's expectations. Having however lost the first three games of the best-of-seven series to the Los Angeles Angels of the Freedom League, the Jays' odds to win shrank to one chance out of a boggle. The boggle is the largest counting number that the human mind can comprehend. Not one person in the whole wide world would bet a single dollar or a yuan for the Jays to win the New World Series.

Yet the Jays did not give up. They fought back one game at a time to tie the series. By winning the seventh and deciding game, the Jays would achieve the unprecedented comeback from three games down to win the New World Series, an impossible feat no team had accomplished in the history of baseball.

Victoria had been trying her best to live a normal life in Toronto while being reconciled to the fact that she was not exactly a member of the Homo sapiens family. As a diehard fan of the Blue Jays, Victoria was glad that they would be hosting the big finale at Rogers Centre, called Skydome long ago when the team first won back-to-back World Series titles in '92 and '93. The home team had the last at bat, which would be all the advantage that the Jays needed. The Jays hadn't won anything significant since the stadium changed its name, but this time Toronto would not be denied.

Tickets were impossible to get, but Victoria didn't need one. She had prepared for this day with a much better plan. She had applied earlier to be an intern for the Blue Jays and was accepted. Her father taught her how to make forceful suggestions. Today, she wore a Blue Jays uniform and worked as their ball girl, picking up dead balls beyond the foul lines.

Anyone who was anybody came to the game. The Rogers board of directors including members of the Rogers family and their longest serving director Phil Lind were in their private box. A large contingent of ex-players, especially those from the World Series years, came to root for the home team. The greatest outfield combo in the 1980s, George 'kiss my purple butt' Bell, 'Canon Arm' Jesse Barfield, and 'Shaker' Lloyd Moseby came. The colour commentator Buck Martinez, the heroic ex-catcher of the Blue Jays who completed an impossible 9-2-7-2 (Barfield to Martinez to Bell to Martinez) double play while sprawled flat on his back with a broken leg and a dislocated ankle wouldn't miss it for the world.

The best second baseman in history Gold Glove Hall of Famer Roberto Alomar, clutch hitting switch hitter Gold Glove shortstop Tony Fernandez, '93 World Series walk off homer hitter hero Joe Carter, and their manager during the glory years Cito Gaston, the first African American manager to win a World Series of the old leagues, joined the opening ceremonies. Carlos Delgado, holder of just about every hitting record of the Toronto Blue Jays came for the festivities. Even Doc Halladay's family came on behalf of the Jays' most beloved pitcher who tragically passed away in a plane accident. There were many others too numerous to mention.

The Angels' new owner the exalted leader Trump and his wife-consort accompanied by Trump's five blue-blooded offspring came to root for their team. Trump acted with extreme hubris but surprisingly gave hope for world peace. He was a paradox and an oxymoron. He was the world's first certified paradoxymoron.

The Trump Empire by now owned most of the valuable corporate assets in America. Trump achieved that by first owning the Fed. The Trump standard sporting the familiar blue trumpet flower Datura, sometimes known as Angel's Trumpet, flew alongside the Angels banners. Seven all-star Angels, each accompanying a Trump royalty, stood at the centre of the diamond when the national anthems were performed for the seventh game. This was at odds with tradition; but what Trump wanted, Trump got. People no longer made a fuss with that phenomenon.

Keanu Reeves and Alex Winter as Bill and Ted sang the Star Spangled Banner, ending the anthem by asking everyone at the stadium and in front of the television to play the air guitar for world peace, closing out their performance with the famous exhortation, "Be excellent to each other and party on dudes."

Shania Twain and Celine Dion belted out the gender-neutral version of *O Canada* in both English and French. The Mounties were there to make sure that all the flags were flying right side up. Blue Jays all-time number one star pitcher Dave Stieb tossed the ceremonial first pitch, a seriously wicked slider, and le jeu commencé—the game begins.

It turned out to be a gem of a pitcher's duel between Angels' golden arm pitcher Jake Arrieta and Jays' top gun Aaron Sanchez.

At the end of the seventh inning, the score was still goose eggs—tied at zero. Then destiny intervened. The inconceivable unpredictable implausible happened.

The fall of the first domino seemed innocuous and unrelated. The organist who had been doing a great job so far had an accident during toilet break. He tripped and broke his wrist. The regular backup was nowhere to be found. The cops had arrested him on his way to the stadium. He had too much alcohol in his blood. The fool started the celebrations early. Now a replacement was urgently needed right away. Someone at the Jays organization knew Victoria played the piano and asked her to take over the organ for the rest of the game.

Victoria would have loved to stay at the field where the action was. If there was going to be a celebration, she wanted to be right in the thick of it. But it wasn't meant to be. Victoria didn't demur; she was the ultimate team player. She settled down at the organist's chair and started to play *Okay Blue Jays Let's Play Ball* with gusto.

Both sides breezed through the eighth inning, but then the top half of the ninth inning went very badly for the Jays. Victoria couldn't bear to watch. The ball decided to take a couple of crazy bounces, a lucky hit way out of the zone, a marginal call, and the Angels was up three-nothing by the time the Jays came out to take their last at bats at the bottom of the ninth. Two outs went by quickly and quietly—a short hop to short and a lazy fly to right field. The atmosphere in the stadium was as gloomy as a morgue. Even the organ lost its heart and sounded like a dirge for a funeral procession. Will the Jays fans be blue for the winter again? Will Toronto Blue Jays give up and meekly go home again? No way, José! Toronto is home.

Baseball is unpredictable; the Uncertainty Principle demands it. With two outs and nobody on, Kevin Pillar kept the game alive with a bloop single into no man's land, ending Arrieta's no-hitter bid. Justin Smoak barely beat the throw to first with a half swing on two strikes. The call was disputed and Smoak was ruled safe after a lengthy review. Next, Randal Grichuk walked after fouling twenty-five pitches. The last pitch was an inside slider that seemed to have brushed his pants. Grichuk immediately trotted to first base to ward off any attempt by the Angels to celebrate. After another lengthy play review, Grichuk got to stay at first base.

Blue Jays fans were kept on tenterhooks because the difference of a fraction of an inch (say, a millimetre) or of a split second would mean that the game was over and the championship lost. Instead, the bases were loaded for whom else but the resurgent José Bautista, the home run champion this year with 72 blasts in 162 games. He also drew the most bases on balls during the regular season at 234, both all-time Hall-of-Famer records.

It was surprising because Bautista returned after retirement, taking gas money just for a chance to play. He told interviewers that he received special training from a mysterious one-eyed Kung Fu master. Everyone had the feeling that destiny was in the air and that Joey Bats would not go quietly tonight. Behind him was Kendrys Morales, who had hit home runs in every playoff game, Edwin Encarnación, who had rejoined the Jays in late season with his big bat, and Russell Martin, the beloved Toronto native, all of them ready, willing, and able to do damage with their bats. No one would be giving up tonight.

The stadium now came alive. The noise level went through the roof and out of the opened dome to reverberate from sea to shining sea. Victoria pounded on the organ with exuberance. What seemed to be impossible half an hour ago now appeared all but certain. Ninth inning baseball is the best embodiment of never giving up.

A few pitches later, the count stood at three balls and one strike. After a long conference at the mound, the private pride of a complete game for the starting pitcher would have to give way to the team's combined efforts to win the championship. Wade Davis, Angel's super closer with an incredible and unprecedented one hundred percent save record this season including the playoffs, trotted out of the bullpen.

Wade Davis had his homerless streak of $125\frac{2}{3}$ innings pitched as a reliever ended by José Bautista in 2015. It was the longest homerless streak by a reliever in the history of baseball, and could have been longer, much longer, if not for Bautista. Strictly by coincidence, Davis had again pitched exactly $125\frac{2}{3}$ innings without giving up a home run. Davis wanted payback. What better place than here? What better time than now? He threatened to blow a couple of wicked pitches pass Bautista, sending the old timer to his retirement again and the Jays packing with a game-over strike out.

As for Bautista, he challenged Davis to make the same mistake again. Bautista set himself up at the plate. He paid attention to the defence on the field, and looked for clues from Davis's pitch action. He regulated his breathing, tuned his mind into a blank, and zoned in on the spin, speed, and path of the ball.

With the music of *Kung Fu Fighting* playing in his head, Bautista's heartbeat slowed. He could hear his heart pounding like a war drum in the cavernous silence of the stadium. Bautista's mind said, "Fast ball, very fast. Make mighty Kung Fu swing if I see a hint of something juicy in the strike zone." Davis's cut fastball came at Bautista like a bullet, but it's just wide. Bautista didn't even flinch. The ball was out of his zone even at the point of release. Joey Bats had taken ball four for a walk, scoring Pillar, and keeping the game alive.

But wait …

"Steer-rike!" shouted the umpire.

The ensuing eruption almost brought the house down. Bautista's boiling blood bubbled; his eyes turned red with fire. He forgot the mental lessons from his one-eyed teacher. He flipped his bat into high heaven and started shouting expletives at the umpire, mostly in Spanish. Cheesy nachos and paper cups filled with beer rained down on the umpires and littered the outfield. Security personnel had to scramble over to restrain hotheads from spilling over from the stands to lynch the umpire. Then the second domino fell and real trouble began.

The umpire threw Bautista out of the game. Jays Manager John Gibbons hurriedly entered the fray. Every Jay from the dugout, the bullpen, and on the field followed the lead and surrounded the umpires. Bautista's teammates and coaches were one hundred and ten percent behind him. If Bautista went down, the whole team would go down with him.

Then the third domino fell; the inconceivable happened. The umpire threw out every single person in a Blue Jays uniform who had walked over to intimidate him. That was everyone, EH-VRY-ONE! The Jays had three runners on base but no one to hit the next pitch. By the time the dust settled, the Jays had basically conceded the game. The officials and the reporters were already noting that the game had ended at 11 p.m.

The Global MLB commissioner who was at the game to hand over the trophy decided to make an eleventh hour concession to placate the fans. The giant video screen showed the following message: "Any member of the Toronto Blue Jays organization, who is in uniform for the game but not involved in the fracas on the field, will be allowed to hit for Bautista." It seemed that the only person who could fit the description was a teenage girl by the name of Victoria Solana.

Victoria walked to the home plate with Bautista's bat. She looked silly wearing Russell Martin's outsized helmet. Meanwhile, the video producer had the presence of mind to put Marilyn Monroe's version of *Diamonds are a Girl's Best Friend* on the giant screen.

Everyone knew that it was just a gesture and a formality to finish the game. No one had thought of the possibility that a teenage girl would be taking the full count pitch at the most crucial juncture of the deciding game of the championship. No one expected a teenage girl who was not a professional athlete to be able to touch Davis's wicked fastball; most major league hitters couldn't. Everyone by this time accepted that the Jays had lost the game and the New World Series. Just let Davis make the pitch and get it over with, so everyone could go home.

Victoria had something else in mind—it's called never giving up. She knew what it meant and the time had arrived for her to be tested.

Victoria loved baseball lore, in particular the one about a seventeen-year-old girl pitcher who faced Babe Ruth and Lou Gehrig in her first professional baseball game. Jackie Mitchell took the mound in the first inning with two on and none out. Both the Sultan of Swat and the Iron Horse were powerful left-handed hitters. Jackie Mitchell was a left-handed pitcher who threw a sidearm sinker. It was a reasonable matchup for an exhibition game. There would be no harm done if the men hit back-to-back home runs as expected of them. Instead, the teenage girl struck out Babe Ruth looking with four pitches, and whiffed Lou Gehrig three times on three pitches to strike him out as well. It proves that there is nothing a girl cannot do if she puts her mind to it.

Taking her place behind the home plate, Victoria raised her hand and pointed at the outfield—a signal that she would hit one out of the stadium—an allusion to Babe Ruth's famous call shot, drawing a ripple of laughter at such a preposterous gesture. Finally, the crowd quieted down. Viola watching the game live in China had her jaws dropped to her knees. Charlie watching the game at Princeton was likewise stunned. Huang Shi was watching the game with his grandchildren at David's home. He had a premonition that something momentous was going to happen. Victoria's real parents, Piero and Luna, had just placed a bet in the Casino Niagara on the Jays to win. They happened to know something the odds-makers didn't. Bella and Jackie had been jumping up and down at home when they saw Victoria at the plate, but now they also calmed down, biting their nails nervously as Davis wound up his pitch.

The time was eleven minutes past eleven. Davis made a note of that. It would be awesome to end this grand baseball duel at this particular moment so that it would be remembered forever. Davis made no mistake. He let loose his signature hundred miles-an-hour four-seam fastball right down the pipe. He did it seamlessly and gracefully. All of his mechanisms were perfect. It was a beautiful thing to watch, a work of art which began as a Michelangelo and ended as a Monet, which gave the impression of speed, like the fleeting feeling of something flying by, but by the time you sensed it, it's no longer there—like the speed of winged thought, like the illusion of time itself.

The ball was a blur. As Victoria concentrated on the pitch, time slowed. She could see the ball spinning as it barrelled through the air towards her. Victoria knew she was strong. She could easily arm-wrestle Davis and Bautista to the ground with one arm, but all she needed to do was to hit the ball on the sweet spot of the bat and follow through with the swing. She had a pretty good idea how to go about doing that. If it's a game, then she's game.

Instead of the thud of a fastball hitting the catcher's mitt and the whiff of a bat hitting air, everyone was surprised to hear a loud crack. The whole stadium and a billion television viewers stared in stunned silence as the ball went up, up, up, and into the fifth deck, measuring at 495 feet (165 metres), a Rogers Centre record.

Victoria looked up to follow the trajectory of the ball, but she saw something else. She didn't hear the celebrations and fireworks exploding around her. She didn't notice the whole team rushing out into the field at her. She didn't even feel the thrill of being carried and paraded around the stadium. What she saw was bad news … very bad news. It's tough luck for Homo sapiens.

Victoria's eagle eyes could see that a red star had blown up in a supernova at this very instant. It was the red giant Betelgeuse, the hand of Orion the Hunter, the solar system of her mother's home planet. Although the supernova happened about six hundred and fifty years ago, around the time when the Ming dynasty was being established by a man who was once a wandering mendicant monk, after peasant rebellions were started by the oracle of a *stone man with one eye*, the news finally reached earth.

In a few years, the alien refugees would start arriving in swarms from the Kuiper Belt, maybe eventually in the tens of millions. The huge supply of Neodymium mined by Xin Feng's clan ensured that. The authorities of St. Paul, Alberta might want to tear down their UFO landing pad and the Aliens Welcome sign. How could all of us be so blind to the numerous signs of the apocalypse staring us in the face?

But seriously, did Victoria really hit an ultimate grand slam to win the New World Series for the hometown Blue Jays? Is this a dream or is this Victoria's fantasy? Is this the latest *Cranial Imaging Immersion Game* or a contrived *Virtual Quantum State Simulation*? Is this in the future or has it already happened? More importantly for us Homo sapiens, has Betelgeuse really exploded and will we as a species survive a large influx of alien refugees? Will Shangala intervene on behalf of earth? If not, can earth defend itself?

Other than using our assault rifles and weapons of messy destruction to terrify and slaughter our own kind, will we finally be able to use them effectively against a much more advanced alien life form? Will our beloved weapons even work on them? Will the aliens do to us what we did to the Neanderthals, Denisovans, Homo erectus, Homo Naledi, and myriad other flora and fauna standing in the way of our domination of the earth? Will they enslave us as we have enslaved our own kind? Will they respect our property rights or will they take away everything we own?

On a more somber note, will they lock us up in tiny dark cages from birth, shoot us full of chemicals, and let us stew in our own excrement so they can cheaply breed us for food? Will they use us as guinea pigs for their new medical and consumer products? Will they perform vivisection on us for testing biological and chemical weapons? Will they massacre us or will they hunt us for sport? Will we become them or will they become us or will we turn into something else altogether?

Is there anything we can do to ensure our survival as human beings—assuming being human is as great as our own lies have been telling ourselves? Will Homo sapiens even leave a trace in the universe, or will we be expunged and be forgotten as if we have never existed?

Alternatively, will the aliens act completely inhuman by disabling all our weapons, sharing their unlimited energy and food production technologies with us, and teaching us how to live in love and peace? These are all questions that need answers before the first wave arrives.

It sounds scary, but don't panic. We have a trump card that we can count on. Victoria is on our side, and no matter who or what will be coming from Betelgeuse, she will tickle their catastrophes, or in other words, kick their collective butt. But that, my friend, is another story.

Image licenses and acknowledgement

Cover image (Rosette Nebula) is licensed under a Creative Commons Attribution Share Alike 3.0 United States License. Created by Adam Block and Tim Puckett at Caelum Observatory (http://www.caelumobservatory.com/). NGC7293 by NASA and ESA (public domain)

Chapter 6, Orion Constellation, by Torsten Bronger under a Creative Commons Attribution Share Alike 3.0 License. (https://commons.wikimedia.org/wiki/File:Orion_constellation_map.svg)

Chapter 9, Scorpius constellation, by Till Credner under a Creative Commons Attribution Share Alike 3.0 License. (https://commons.wikimedia.org/wiki/File:ScorpiusCC.jpg)

Chapter 14, Looting of the Old Summer Palace during the Second Opium War (1860) by Godefroy Durand (public domain)

Chapter 15, Dante's Inferno: Titans and giants by Gustave Doré (public domain)

Chapter 21, Money cowry by H. Zell under a Creative Commons Attribution Share Alike 3.0 License. (https://commons.wikimedia.org/wiki/File:Monetaria_moneta_01.JPG)

Chapter 21, Square shoulder spade coin by David Hartill under a Creative Commons Attribution Share Alike 3.0 License. (https://commons.wikimedia.org/wiki/File:Square_Shoulder_Spade.jpg)

Chapter 22, Revelation: Opening of the Sixth Seal by Albrecht Dürer (public domain)

Chapter 25, Plastron and Scapula by Babelstone under a Creative Commons Attribution Share Alike 3.0 License. (https://commons.wikimedia.org/wiki/File:Shang_dynasty_inscribed_tortoise_plastron.jpg) (https://commons.wikimedia.org/wiki/File:Shang_dynasty_inscribed_scapula.jpg)

Chapter 29, Shang Ge (Bronze dagger-axe) by a Wikipedia editor at large under a Creative Commons Attribution Share Alike 2.5 License. (https://commons.wikimedia.org/wiki/File:CMOC_Treasures_of_Ancient_China_exhibit_-_bronze_dagger-axe.jpg)

Chapter 39, Chinese unicorn Qilin image from Ming dynasty (public domain)

Chapter 53, Tomb of Fu Hao by Chris Gyford under a Creative Commons Attribution Share Alike 3.0 License. (https://commons.wikimedia.org/wiki/File:Tomb_Fu_Hao_YinXu.jpg)

Chapter 59, Osiris hieroglyph (public domain)

Chapter 59, The Eye of Horus by Jeff Dahl under a Creative Commons Attribution Share Alike 4.0 License. (https://commons.wikimedia.org/wiki/File:Eye_of_Horus_bw.svg)

All other graphics created by the author.

CPSIA information can be obtained
at www.ICGtesting.com
Printed in the USA
LVHW04s0028131018
593503LV00001B/68/P

9 781999 401900